Praise for
The Homeowner's Guide to Renewable Energy

If you're thinking about investing in a renewable energy system for your home,
Dan Chiras' *The Homeowner's Guide to Renewable Resources* helps clarify the
decision-making process. After beginning with the all important discussion about
energy efficiency and conservation, Dan guides you through everything you need
to choose which renewable options to integrate into your lifestyle.
A great addition to my bookshelf!

— Mick Sagrillo, Sagrillo Power & Light

In 1975, when we first went off the grid, there was, at best, only sketchy
information on wind power; photovoltaics were just starting to be heard from;
and small hydro was a rarity. Equipment was limited and knowledge of how to use it
was scarce. If we'd had a Dan Chiras all those years ago, oh, the mistakes we could have
avoided! Thankfully, those who want to go off-the-grid today — or even interface with
the grid using their home-grown power — have Dan's clearly written and up-to-date
Guide to all forms of renewable energy to help them make informed choices.
It's all from a guy who has been there, done that.

— Rob Roy, Director, Earthwood Building School and author of
Earth-Sheltered Houses and several other books on green building

Dan Chiras is one of the most authoritative writers in the field of
renewable energy. His multiple other books create a comprehensive library for
homeowners looking to live a lifestyle in harmony with their values. Not only is his
style accessible and easy to read but is thorough in what to do, how to do it and why.
Dan walks his talk living in a solar, green home and devoting untold hours
to sustainable living causes. He is truly one or our national heroes!!!

— David Johnston, author of *Green Remodeling* and
Green From the Ground Up

Who says home energy improvements have to be complicated, or boring? Dan Chiras' *The Homeowner's Guide to Renewable Energy* waltzes the reader gracefully through various efficiency upgrades that put household heat and coolness in their proper places, then expertly jazz-dances through state-of-the-art technologies like solar electricity, heat pumps, and biodiesel fuel. This easy-to-understand, timely book should be distributed by local governments and utilities to homeowners throughout the country. Imagine how much less dependent we'd be on uncertain, expensive supplies of oil and natural gas!

— Dave Wann, coauthor of *Affluenza: The All-Consuming Epidemic* and *Superbia! 31 Ways to Create Sustainable Neighborhoods*

Dan Chiras strikes again! With this latest addition to his already impressive list of titles, Dan makes it as easy as possible for you to effect your own transition away from fossil fuel dependence. I've never seen a more comprehensive, better written, or better organized primer on this subject. When you need practical advice from a warm, smart and informed human being, Dan Chiras is the one to turn to.

— Bruce King, PE Director, Ecological Building Network, and author of *Buildings of Earth and Straw* and *Making Better Concrete*

The Homeowner's Guide to Renewable Energy makes abundantly clear the predicament that humankind has created regarding how we procure and use energy. Ways that we might extricate ourselves from this predicament are placed squarely on the shoulders of renewable forms of energy, rather than fossil fuels. This book shows how we, as individuals, have the power and technology available now to embrace renewable energy for a bright future.

— Kelly Hart, www.greenhomebuilding.com

REVISED & UPDATED EDITION

THE
HOMEOWNER'S GUIDE
TO RENEWABLE
ENERGY

ACHIEVING ENERGY INDEPENDENCE THROUGH
SOLAR, WIND, BIOMASS and HYDROPOWER

DAN CHIRAS

NEW SOCIETY PUBLISHERS

Cover design by Diane McIntosh. © iStock (RapidEye)

Printed in Canada.
First printing June 2011.

Paperback ISBN: 978-0-86571-686-5
eISBN: 978-1-55092-391-9

Inquiries regarding requests to reprint all or part of
The Homeowner's Guide to Renewable Energy – Revised & Updated Edition
should be addressed to New Society Publishers at the address below.

To order directly from the publishers, please call toll-free
(North America) 1-800-567-6772, or order online at www.newsociety.com

Any other inquiries can be directed by mail to:

New Society Publishers
P.O. Box 189, Gabriola Island, BC V0R 1X0, Canada
(250) 247-9737

Library and Archives Canada Cataloguing in Publication

Chiras, Daniel D.
    The homeowner's guide to renewable energy : achieving energy independence through solar,
wind, biomass, and hydropower / Dan Chiras. -- Rev. and updated ed.

Includes index.
ISBN 978-0-86571-686-5

    1. Dwellings--Energy conservation. 2. Renewable energy sources. I. Title.

TJ163.5.D86C48 2011   644   C2011-902967-7

New Society Publishers' mission is to publish books that contribute in fundamental ways to building
an ecologically sustainable and just society, and to do so with the least possible impact on the environment, in a manner that models this vision. We are committed to doing this not just through education,
but through action. Our printed, bound books are printed on Forest Stewardship Council-certified
acid-free paper that is **100% post-consumer recycled** (100% old growth forest-free), processed chlorine free, and printed with vegetable-based, low-VOC inks, with covers produced using FSC-certified
stock. New Society also works to reduce its carbon footprint, and purchases carbon offsets based on
an annual audit to ensure a carbon neutral footprint. For further information, or to browse our full list
of books and purchase securely, visit our website at: www.newsociety.com

NEW SOCIETY PUBLISHERS

MIX
Paper from
responsible sources
FSC® C016245

Dedication

This book is dedicated to the renewable energy and energy-efficiency pioneers throughout the world, like my good friends Henry Rentz of Missouri Valley Renewable Energy in Hermann, Missouri, small wind expert Mick Sagrillo of Sagrillo Power and Light in Wisconsin, and Bob Solger at Brightergy — and hundreds like them from Maine to Montana — who have helped make the shift to a renewable energy economy possible, often at great personal expense.

This book is also dedicated to the founders and the hard-working staff at *Mother Earth News*, *Home Power* magazine, and *Solar Today*, and to the staff and volunteers of scores of nonprofit groups such as Solar Energy International, the Colorado Renewable Energy Society, The American Solar Energy Society, the American Wind Energy Association, the Midwest Renewable Energy Association, Heartland Renewable Energy Society, Show Me Solar, the Iowa Renewable Energy Association, the Solar Living Institute, the National Renewable Energy Laboratory, the Great Lakes Renewable Energy Association, all of the remaining chapters of the American Solar Energy Association, and last but far from least, the Institute for Sustainable Energy Education.

My deepest thanks to all of them for their dedication, hard work, and perseverance — and for helping create a more sustainable energy system that promises a cleaner, healthier, more economically viable future for all of us.

Books for Wiser Living
recommended by *Mother Earth News*

Today, more than ever before, our society is seeking ways to live more conscientiously. To help bring you the very best inspiration and information about greener, more sustainable lifestyles, *Mother Earth News* is recommending select New Society Publishers' books to its readers. For more than 30 years, *Mother Earth* has been North America's "Original Guide to Living Wisely," creating books and magazines for people with a passion for self-reliance and a desire to live in harmony with nature. Across the countryside and in our cities, New Society Publishers and *Mother Earth* are leading the way to a wiser, more sustainable world.

Contents

Acknowledgments xi

Introduction 1
   Renewable Energy and Me 3
   Organization of the Book 6

Chapter 1: Renewable Energy — Clean, Affordable, and Reliable 9
   Making Wise Choices 11
   Understanding Energy 14
   What is Renewable Energy? 21
   The Pros and Cons of Renewable Energy 22
   Prospects for the Future.. 27

Chapter 2: Conservation Rules —
   The Cornerstone of Your Energy Future... 31
   Efficiency First!... 33
   Energy Conservation as a Renewable Source of Energy 38
   Benefits of Energy Conservation.. 39
   Home Energy Use 40
   Retrofitting Your Home for Energy Efficiency 42
   The Silver Lining.. 72

Chapter 3: Solar Hot Water Systems —
Satisfying Domestic Hot Water Needs with Solar Energy 75
    Conventional Hot Water Systems 75
    Tankless Water Heaters 78
    What is a Solar Hot Water System? 81
    A Brief History of Solar Hot Water 81
    Solar Hot Water Systems 84
    Which System is Best for You? 100
    Sizing Your System 101
    Finding a Competent Installer 104
    The Economics of Domestic Solar Hot Water Systems 105

Chapter 4: Free Heat — Passive Solar and Heat Pumps 109
    What is Passive Solar Heating? 110
    Is Passive Solar for You? 112
    Types of Passive Solar Design 115
    Getting the Help You Need 124
    Some Final Thoughts on Passive Solar Retrofits 125
    Heat Pumps 126
    Conclusion 129

Chapter 5: Solar Hot Air and Hot Water Systems:
Affordable Heat from the Sun 131
    Solar Hot Air Heating Systems 131
    Solar Hot Water Heating Systems 142
    Solar Hot Water Space-Heating Systems 143
    Conclusion 148

Chapter 6: Wood Heat 149
    Retrofitting Fireplaces for Efficiency 149
    Fuel-Efficient Wood Burning Stoves 150
    Shopping for an Efficient, Clean-Burning Wood Stove 154
    Pros and Cons of Wood Stoves 158
    Wood Furnaces 162
    Pellet Stoves 164
    Masonry Heaters 166

Chapter 7: Passive Cooling — Staying Cool Naturally................................173
  What is Passive Cooling?..174
  Tools in the Passive Cooling Toolbox..175
  Reducing Internal Heat Gain..175
  Reducing External Heat Gain..176
  Building a Better Future..195

Chapter 8: Solar Electricity — Powering Your Home with Solar Energy 197
  What is a Solar Electric System? 200
  Buying a Solar Electric System..223
  Locating a Reliable Contractor..237
  Why Install Solar Electricity?..238

Chapter 9: Wind Power — Meeting Your Needs for Electricity............................239
  Is Wind Power in Your Future?..240
  Wind Power: A Brief History..241
  Understanding Wind Generators 243
  Wind Systems: Three Basic Options..246
  Is Wind Energy Appropriate Where You Live?................................247
  Selecting a Wind Generator and Tower..249
  Financial Matters..262
  Wind Power without Installing a Wind Generator..........................263

Chapter 10: Microhydro — Generating Electricity from Running Water............267
  An Introduction to Hydroelectric Systems 268
  The Anatomy of a Microhydro System 269
  Assessing the Feasibility of Your Site 271
  Assessing the Potential of a Microhydro Site 273
  Buying and Installing a System 281
  Finding an Installer or Installing a System Yourself 290
  The Pros and Cons of Microhydro Systems 290

A Brief Afterword 293
Appendix..295
Resource Guide..297
Index 327
About the Author..236

JOIN THE CONVERSATION

Visit our online book club at www.newsociety.com
to share your thoughts about *The Homeowner's Guide to Renewable Energy*.
Exchange ideas with other readers, post questions for the author,
respond to one of the sample questions or start your own
discussion topics. See you there!

Acknowledgments

This book has been made possible by many other hard-working and dedicated individuals who have spent a lifetime exploring, teaching, installing, and writing about energy efficiency and renewable energy. Many of their names and the names of the organizations they've helped to build grace the pages of this book, especially in the Resource Guide at the end of the book. Without them, this book never could have been possible. Without them, renewable energy would still be a wishful dream. So, a world of thanks to all of you! Keep up the hard work, and thanks a million for your efforts.

I am also deeply grateful to the people who answered my many questions on renewable energy and energy efficiency when writing and revising this book, among them Jeff Scott, Henry Rentz, Mick Sagrillo, Ian Woofenden, Johnny Weiss, Randy Udall, Marc Franke, and Steve Andrews. A special thanks to Johnny Weiss, Marc Franke, Dan New, Mick Sagrillo, and Randy Udall for reading portions of the first edition manuscript, offering helpful comments and advice, and helping to ferret out inadvertent mistakes. Many thanks to those who provided the photographs for this book.

I would also like to express my thanks to the many generous individuals who helped me establish The Evergreen Institute Center for Renewable Energy and Green Building, my educational center in Gerald, Missouri (evergreeninstitute.org). There, I have gained considerable knowledge about renewable energy and honed my skills. My teaching at The Evergreen Institute and through other organizations has opened my eyes to a wealth of new information, much of which has been incorporated into this new edition of *The Homeowner's Guide to Renewable Energy*.

First and foremost among those who helped me build a world-class energy and green building education center is my friend Rocky Huffman at Kansas-based Primo Sustainable Products and Sustainable Energy Products. Rocky super-insulated our 2,400 square foot classroom building and our staff headquarters; he also donated many buckets of lime putty, which we used to plaster our straw bale cottage and parts of the classroom. Rocky's family salsa (Momma Salsa) made its way into many after-class parties, too.

Many thanks also to local excavator Dan Boman of Custom Power Excavating who performed countless valuable tasks, including all of the excavation for our solar electric and wind energy systems and our straw bale building. He also helped weld the frames for our solar electric systems. And he even helped us with our erosion control efforts. Dan has been an extremely valuable ally.

Thanks also to my friend Jeff Scott of Sol Source in Denver who donated solar modules and a solar attic fan I use for demonstrations. Jeff supplied one of our two solar electric systems that today help

The Evergreen Institute produce 100% of its power.

Many thanks to my brother Jim who helped remodel our classroom building — for example, the new wiring and framing that allowed us to super-insulate the building. Jim also donated office equipment and lots more!

Many thanks to Don Cary who donated countless hours to help remodel our classroom building. Don did nearly all the drywall, mudding, and taping. A world of thanks to my friend and confidante, Tom Bruns, an engineer who helped install tile and pipe, gave advice on foundation design for our newest PV system, and helped design and build the roof for our straw bale cottage where students can now stay while taking classes. Tom continues to be a valuable asset to our center and I look forward to many years of productive collaboration.

Many thanks go to Mike Beerbower who traded his time to help us build the straw bale cottage, tile the floor in the classroom, and much more. A world of thanks go to Henry Rentz of Missouri Valley Renewable Energy who helped install our wind system foundation and two solar hot water systems, and who has provided advice on our newest solar electric system and helped any time we needed assistance. Many thanks to Bob Solger, former owner of the Energy Savings Store, who donated our Skystream 3.7 Wind Turbine and to Joe Steenbergen of Victor Energy who traded the tower pipe for our 126-foot tilt-up guyed tower. Many of these folks — and others, like Pete Veronesi and Mark Schueneman — gave freely of their time or traded services and materials for training.

Thanks also to all the students who helped install our various renewable energy systems during workshops and who helped build our straw bale cottage and remodel our classroom building.

Finally, I would also like to express my appreciation to my dearest friends Chris and Judith Plant at New Society Publishers who signed on to this book many years ago and have remained cordial, enthusiastic, and supportive throughout the revision and production of this book, the revised edition, and all of the other books I've written for their amazing company.

I would also like to thank all of the dedicated staff at New Society Publishers, including my friend Ingrid Witvoet; copyeditor Linda Glass, for her thoughtful and skilled copyediting; Jill Haras and Anil Rao, Ph.D. for their excellent drawings; Sue Custance for handling the countless production details; Greg Green for his expert design and layout; and Sara Reeves for her considerable efforts to publicize this book.

Introduction

For years, Kara Culpepper and her family held season tickets to the Denver Broncos, attending every home game, like tens of thousands of avid fans. Those were good times, despite the fact that the family often had to endure some of Colorado's most bitterly cold winter weather to watch their favorite team. But after taxpayers built the team a new stadium and ticket prices went through the roof, Kara and her family decided to give up their season tickets. It was just too costly. They didn't give up on their team, however. They decided to watch games on TV at home. Trouble is, their modest suburban home, which was built in the 1970s, was an icebox in the winter. Their home was woefully under-insulated and full of leaks that allowed cold air in on blustery winter days — like

millions of similar residences throughout North America, indeed the world. To watch the games, Kara and her family had to bundle up in jackets and sweaters or huddle under blankets. Ironically, it wasn't a whole lot different than a cold December game at the stadium. "The only difference was that in the stadium, you could actually get sun rays," remarks Kara. So they were a bit colder at home.

Fortunately, those days have ended. Today, Kara and her family watch the Broncos games in comfort, no longer bundled up like the Inuit on a cold Arctic night. Their home is no longer just bearable, it has become quite comfortable thanks to an extensive home energy retrofit. The energy retrofit was made possible by a generous grant from their local utility (Xcel Energy) and a nonprofit organization, the Colorado Energy Science Center, a leader in wise energy use in Colorado.

This all happened because Kara and her family qualified for a complete energy makeover, worth over $25,000. Her home was selected as one of two winners in a statewide competition. Among an applicant pool of 10,000, her home was judged to be one of the two most energy-inefficient homes. The judges believed that retrofitting her home would provide significant energy and cost savings. Being voted as owners of one of the most energy-inefficient homes in the state of Colorado is not a great distinction, but Kara and her family were able to look past that dubious honor. The extensive energy retrofit they received has dramatically cut their heating bills and increased their comfort levels beyond their wildest imaginations.

In short order, the family witnessed dramatic changes. Air sealing, improved insulation, and a host of other energy upgrades slashed their natural gas bill in half, saving the family $150 per month in the dead of winter. New energy-efficient appliances that replaced older, less frugal models, saved the family hundreds of dollars a year on their electrical bills.

Besides saving energy and money, their home is now much more comfortable — a priceless benefit. Their home is much warmer in the winter and much cooler in the blisteringly hot Colorado summers. All told, the energy retrofit has reduced the family's emissions of carbon dioxide by about eight tons a year! "That's roughly equivalent to removing one and a third vehicles from the highway every year," writes Amanda Leigh Haag in *Smart Energy Living*.

If you are like most of your neighbors, you're being hammered by high fuel bills — at home, at your business, and at the gas pump — and you want to do something about it. Like Kara Culpepper's family, you can reduce your energy consumption dramatically — but you don't have to pay $25,000 to do so! An investment of three or four hundred dollars, in fact, can result in amazing energy savings that are good for your wallet and good for your future — as well as the future of your children and theirs, and the many species that share this planet with us. A few thousand dollars will bring even greater benefits!

You can increase your energy independence by joining the growing number of homeowners and business owners in urban, suburban, and rural settings the world over who are not only using energy more efficiently, but also producing some — or in many cases all — of their own energy from renewable resources like the sun or wind. With knowledge, careful planning, and a little money, you can also free yourself from ever-rising fuel bills by turning to efficiency and clean, reliable, and affordable renewable energy technologies. If this is your dream, this book is for you. The previous edition has sold about 40,000 copies to date. Like its predecessor, this edition will help you pursue your dreams of greater energy self-sufficiency and a comfortable and affordable life.

The revised edition of this book will, first and foremost, help you understand *all* of the renewable energy options at your disposal. It will help you develop a sensible, cost-effective strategy to slash energy use and increase your reliance on renewable energy.

RENEWABLE ENERGY AND ME

I have a long-standing love affair with renewable energy. In fact, I fell in love with this clean alternative to mainstream energy in the summer of 1977 while visiting Arches National Park in Moab, Utah. It all occurred in the most unlikely spot — in the parking lot in front of the visitors center. There, park officials had placed the single solar electric module shown in Figure 1.

In the baking hot summer sun, this amazing little device was cranking out electricity that powered a small fan. Park officials had attached streamers to the fan to dramatize the effect. My immediate interest in this amazing, quiet device was sparked partly because I'd been studying

the impacts of generating electricity from conventional sources, notably coal and nuclear fuels. I'd heard about solar electricity and seen pictures of various solar technologies, but had never seen a solar electric module in operation. And there it was, this elegantly simple alternative to messy fossil fuels and dangerous nuclear

Fig. 1: *This small display of solar electricity turned my head and started a lifelong commitment to renewable energy. Unfortunately, you can't see the fan and streamers that dramatized the PV's remarkable ability to convert solar energy into electricity.*

DAN CHIRAS

Fig. 2: *A PV array at The Evergreen Institute provides the electricity that powers a classroom and the author's office. Similar systems could be used to power homes throughout the world, providing clean, renewable energy for decades.*

power plants. It struck me as a perfect way to live in harmony with nature.

I remember thinking that if solar electric modules like these were placed on millions of roofs throughout North America, they could power the entire world (Figure 2). Sure, it would require resources to make solar systems, but a lifetime of free fuel — solar energy — certainly makes this a much better resource than conventional fuels.

Today, sitting in my wind-turbine-and-solar-module-powered office, I'm living my dream. Out in the field next to my office is a quiet solar array, gleaming in the bright sun, converting solar energy into electricity. I never cease to marvel at the fact that the energy that's powering my lights and my computer is solar energy that jetted

through space, covering 93 million miles in 8.3 minutes. My simple, reliable PV array has no moving parts — like the module I first encountered in Moab, which was cranking out electrical energy in the hot desert sun and sending those streamers and my heart into paroxysms of delight.

Since that day, I have devoted my life to the study of renewable energy, including solar energy, wind power, hydropower, geothermal energy, tidal power, biofuels, and hydrogen. I have written about renewable energy in several books, including college textbooks. I published a bestselling book titled *The Solar House* that describes how to heat and cool a home passively — without costly heating and air conditioning systems and the polluting fossil fuels or dangerous nuclear fuels that power them. My other books include *Power from the Wind, Power from the Sun, Wind Power Basics, Solar Electricity Basics, Solar Home Heating Basics,* and a book on sustainable transportation, *Green Transportation Basics.*

I've done more than research and write about the many potentially liberating renewable energy technologies; I have put my knowledge into practice. In the 1970s, I retrofitted my very first home with solar hot water panels for domestic hot water, virtually eliminating my hot water bill. I packed my attic full of insulation to save energy and built a small solar greenhouse on the house's south side to warm it. I installed a wood stove (and gathered wood for free from a nearby national forest). Together, the insulation, greenhouse, and wood stove virtually eliminated my heating bill.

My second home, purchased many years later, was a passive solar house. Although it worked pretty well, I found ways to improve its energy performance, reduce my family's energy bills, and achieve greater self-sufficiency. In 1995, I built a super-efficient solar home from scratch. This energy-miserly house generates 100 percent of its electricity from a solar electric system and small wind generator, freeing me from those nagging monthly utility bills (Figure 3). (I haven't paid an electrical bill since 1996!) I also have the satisfaction of knowing that I'm dramatically reducing my family's impact on the environment. Our home is passively heated by the sun through south-facing windows, and it is cooled naturally as well. I burn a cord of wood a year as backup heat; all in all, it costs me about $150 per year to supplement the sun's free heat. I plan to install a solar hot water system that will heat the house, which will eliminate the need for the wood stove.

I have no air conditioner. I don't need one. The house stays cool through the hot summer months thanks to high levels of insulation, energy-efficient windows,

earth sheltering, and other features I'll explain later in this book.

In the fall of 2008, I purchased a 50-acre farm in Missouri, which has been converted to an educational center, The Evergreen Institute Center for Renewable Energy and Green Building. Here, I teach numerous classes on all sorts of renewable energy systems and green building. The classroom and faculty residence have been retrofitted to make them super-efficient; they are equipped with two solar hot water systems, two solar electric systems, and a wind turbine on a 126-foot tower. We have even converted a small pickup truck to electricity. Our goal is to produce as much energy as we consume.

You too can dramatically reduce your energy use. You can even achieve near total

Fig. 3: *The author's passive solar/ solar electric home in Evergreen, Colorado obtains energy from a solar electric system and a small wind generator.*

energy independence, eliminating the sting of high monthly fuel bills and greatly reducing your environmental impact. This book will show you how. I'm not going to dwell on the world's energy problems — peak oil and natural gas, the high cost of fuels to our economy, the plethora of environmental consequences of conventional fuel use, global climate change, etc. I discussed those in the previous edition. With this edition, I'm going to push forward, immediately delving into clean, reliable, and affordable solutions that you can use to reduce energy consumption and gain greater self-sufficiency at home and in your own business.

ORGANIZATION OF THE BOOK

My goal with this book is to show you that it is possible for you to incorporate renewable energy into your daily life. Chapter 1 will give you an overview of renewable energy, examining the many options available to homeowners. I also present a summary of the pros and cons of renewable energy, so you can enter into this venture with eyes wide open.

In Chapter 2, I'll explore efficiency and conservation, the cornerstones of personal energy strategies. You will see why I consider energy conservation and efficiency efforts to be extremely valuable forms of renewable energy. You will also see that you can save lots of money through

home energy efficiency and conservation efforts, and that there are a host of other benefits. This chapter concludes with a simple, cost-effective home energy efficiency and conservation strategy that will save you and your family thousands of dollars, perhaps tens of thousands of dollars, over your lifetime. These ideas can even be applied to your business.

In Chapter 3, we'll focus our attention on solar hot water systems for providing domestic hot water. We will examine the types of systems on the market today, and explore how they work. I'll provide information, including costs, that will help you decide which system is best for your home. As in other chapters, we'll explore important home energy savings that will reduce your demand for hot water. We'll also look at the pros and cons of solar hot water systems.

In Chapter 4, I explore some solar space-heating options, that is, how you heat your home and business using renewable energy. I discuss one of my favorite strategies, passive solar heating: heating homes without costly nonrenewable fuels or expensive mechanical systems. I'll also look at a popular technology referred to as heat pumps; these could become a mainstay of American home heating in the not-too-distant future.

Two technologies that you will be seeing more of in the future, covered in

Chapter 5, are solar hot water and solar hot air systems used to heat homes.

In Chapter 6, I will examine wood burning as a home heating strategy. Fireplace inserts, wood stoves, pellet stoves, and masonry heaters are covered in this chapter. I'll also tell you where you can find lots of free wood, even in cities and towns.

Chapter 7 tackles the enormous challenge of cooling a home without costly fossil fuels. I offer some general strategies as well as specific tools of the trade, and I explain how they can be applied to different climate zones.

In Chapter 8, I will explore solar electricity, an amazing technology that converts sunlight energy into electricity. I will describe the options that are available to you as well as the costs of these systems and ways to help offset the costs, sometimes substantially, through efficiency measures and local, state, and national incentives.

How you can produce electricity from wind power is examined in Chapter 9. Although wind generators are not for everyone — certainly not those who live in urban or suburban neighborhoods — you will see that there are even ways that you can tap into wind energy *without* installing a wind generator in your back yard.

In Chapter 10, I explore microhydroelectric systems, a technology that allows rural residents in some areas to tap into the power of flowing water. Although few residential sites in North America are appropriate for this approach, those who are lucky enough to live on one will find microhydroelectric to be an excellent choice.

At the end of the book is a list of important resources — websites, books, articles, magazines, videos, organizations, and so on — that can provide additional information and support.

CHAPTER 1

RENEWABLE ENERGY

CLEAN, AFFORDABLE, AND RELIABLE

Contrary to what many people think, renewable energy is not a source of energy we've just discovered. Humans have relied on renewable energy since the very first humanlike creatures roamed the planet over three million years ago. Throughout most of human history, the energy human beings needed to survive and prosper has come from food molecules — primarily seeds, berries, and roots. The energy in these foods provided the means by which we built early civilizations. Our early ancestors also burned wood to warm their caves and cook their food.

Plants, of course, are renewable resources, capable of regenerating themselves from seeds, roots, or tubers. But plants are here by the grace of three other renewable environmental resources: soil, water, and air.

Although our predecessors, and virtually all other life forms on the planet, received the energy they needed to survive from plant matter, the source of the energy extracted from our botanical companions is not the soil or water or even the air. The source is the sun — a massive hydrogen fusion reactor 93 million miles from planet Earth.

Plants capture the sun's energy during photosynthesis. In this complex set of chemical reactions, plants synthesize a wide variety of food molecules from three basic "ingredients": carbon dioxide from the air, water from the soil, and solar energy from the sun. Solar energy that drives photosynthetic reactions is captured and stored in the chemical bonds of organic food molecules. When food molecules are consumed by us, or any other animal for that matter, stored solar energy is released. Solar energy contained in food molecules and liberated by the cells of our bodies is, in turn, used to transport molecules across cell membranes and to manufacture protein and DNA to power our muscles and heat our bodies.

Humankind's greatest achievements were made by using the sun's energy. The Egyptians, for instance, hauled massive stones to build the towering pyramids with nothing but ingenuity and the muscle power of conscripted laborers fueled by organic food molecules courtesy of the sun and plants. The Romans expanded their holdings to build a vast and prosperous empire, too, all with horse and human muscle powered by plant matter and, ultimately, sunlight.

For most of human history, then, renewable energy reigned supreme.

Then came the fossil fuel era.

Lumbering to a start in the 1700s in Europe and the 1800s in North America, the fossil fuel era was first powered by coal, an organic sedimentary rock. Coal owes its origin to plants that grew in the Carboniferous era some 250 to 350 million years ago. Coal replaced waning supplies of wood in Europe and fed the industrial machinery that made mass production —

and modern society — possible. So, in a way, the Industrial Revolution was also powered by solar energy — ancient sunlight that was captured by plants millions of years ago.

For many years, coal reigned supreme. But eventually coal was forced to share its kingdom with two additional fossil fuels: oil and natural gas. Also produced from once-living organisms (notably, aquatic algae), these fuels were relatively easy to transport and, like coal, are found in highly concentrated deposits. Over time, oil and natural gas, along with coal, became major components of the world's energy economy.

In 2009 (the latest year for which data were available), oil supplied 37 percent of the United States' total annual energy demand. Natural gas provided about 25 percent, and coal supplied 21 percent of our energy needs. Nuclear energy provided just under 9 percent. The remaining 8 percent of the United States' energy diet was supplied by four renewable resources: hydropower, solar, wind, and geothermal. Canada is similarly heavily dependent on oil, natural gas, and coal. In 2008, oil, natural gas, and coal provided 66 percent of Canada's energy. Nuclear energy provided about 7 percent, and hydropower provided 25 percent of the nation's energy.

In the more developed countries, fossil fuels clearly dominate the energy scene today, but their glory days are coming to an end. Oil and natural gas are entering their sunset years, making the shift to clean, affordable, reliable, and abundant renewable energy inevitable. Fortunately, we have lots of options.

MAKING WISE CHOICES

To make the wisest choices as individuals — and as a society — we need to understand our predicament — what energy resources are endangered. Many energy experts believe that global oil production will peak or already has peaked. Peak oil could result in a devastating rise in prices. Global natural gas production may also peak soon, creating further turmoil. Clearly, we need replacements for these two fossil fuels. But what about coal?

Given the devastating impact and high cost of global warming and a host of other energy-related environmental problems, coal will very likely need to be phased out in the near future. Although coal is abundant in North America, China, and elsewhere, and its use is bound to increase dramatically as oil and natural gas production peak, coal is the dirtiest of all fossil fuels. Coal combustion not only produces sulfur oxides and nitrogen oxides that react with water and sunlight to form sulfuric and nitric acids that poison rain and snow, coal combustion also generates millions of tons of particulates that cause asthma and other respiratory diseases. Coal combustion also

yields millions of tons of ash containing an assortment of potentially toxic materials such as mercury. Much of this ash is disposed of in ordinary landfills alongside our trash, and the toxic chemicals in the ash can eventually seep into ground water. Perhaps most important to our future, however, is that coal combustion produces enormous quantities of the greenhouse gas carbon dioxide — far more carbon dioxide per unit of energy produced than any other fossil fuel in use today (Figure 1-1).

Despite industry's frequent mention of an elusive "clean coal technology," it's difficult to make coal clean. Without question, the efficiency of coal combustion can be increased to reduce the amount of pollution per unit of energy produced, and I applaud any efforts to do so. But capturing

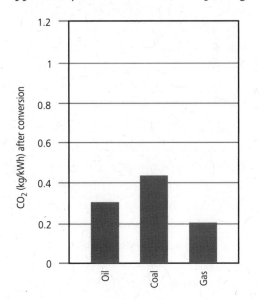

Fig. 1-1: *Not all fossil fuels are created equal. Coal has a much higher ratio of carbon to hydrogen than oil, which has a much higher ratio of carbon to hydrogen than natural gas. The more carbon a fuel contains, the more carbon dioxide it produces per BTU of energy released.*

carbon dioxide and storing it underground — one way to make coal cleaner — is energy intensive and will dramatically increase coal combustion itself. (Carbon capture increases the energy consumption at a power plant by around 25 to 30 percent.) Even with the best technologies in place, coal combustion will produce lots of pollution; it is inevitable, and the more coal we consume, the greater the output of potentially harmful gases and particulates and solid waste. Carbon dioxide, for example, is the unavoidable byproduct of combustion of any carbon fuel. Much of the sulfur that contaminates coal in varying degrees can be removed before or after combustion by pollution control devices. The sulfur, however, does not magically disappear. Most of what is removed by smokestack scrubbers ends up in a toxic slurry that is disposed of in landfills where the toxic components can leach into groundwater. It's a simple mass balance phenomenon: if the chemical ingredients of the pollutants are in the fuel, they're going to be a byproduct one way or another. They won't mysteriously vanish because a coal executive tells you they do. In the end, "clean coal" seems like nothing more than just a deceptive marketing ploy of the coal companies to make a dirty fuel appear more environmentally acceptable.

To make the wisest choices, we also need to understand the end uses of each of the fuels we are trying to replace.

Remember, it is the products and services these resources provide that we want, not the fuels themselves. As natural gas supplies decline, we don't necessarily need more natural gas. We need to ensure the services that natural gas currently provides. For example, many homeowners use natural gas to provide space heat, to heat water for showers, dishwashing, and laundry, and to cook food. Finding replacements for natural gas means finding ways to provide these *services* via clean, renewable resources and technologies — for example, solar hot air systems to heat our homes and solar water systems to provide space heat and heat water for domestic use.

This book lays out the options available to us that can ensure the continuation of services currently supplied by now-failing or environmentally unacceptable fuel sources. Don't forget, though, that the easiest way to meet our needs is often achieved by simply being more efficient. A warm home or business, for instance, can be achieved by sealing up those obnoxious cracks around windows, doors, and elsewhere. It can also be ensured by installing additional insulation to the ceilings and walls of our homes and offices. Additional space heat can be provided by retrofitting our homes for passive solar — adding windows on the south sides of our homes to let in the low-angled winter sun. Space heat can also be achieved by installing active solar hot water systems (Figure 1-2). Solar hot water systems generate hot water that can be integrated with heating systems already in many of our homes — from baseboard hot water systems to radiant-floor systems to forced-air systems. There are many other options out there. For example, homes can be heated via heat pumps, devices that remove heat from the ground or even the air and transfer it to the interior of a building.

To make wise choices, you also need to know which options make the most sense. How do we assess the appropriateness of a renewable energy option?

Two of the most important criteria are cost and net energy yield, which often go hand in hand. Consider an example:

To replace declining supplies of oil, many fossil fuel advocates suggest that we can turn to oil shale and tar sands. Unfortunately, a huge amount of energy is required to extract the oil from these natural resources. The energy required to extract oil from tar sands and oil shales subtracted from the energy of the final product is known as the *net energy yield*. It can be thought of as the energy returned on energy invested. Both oil shale and tar sand oil production have very low net energy yields compared to conventional oil (although new processes have steadily improved the net energy yield of tar sand

production). The lower the net energy yield, the more costly the fuel. As the price of conventional oil increases, the cost of oil shale and tar sand oil will inevitably rise.

Environmental impact should also be a key criterion when selecting an alternative fuel. To build a sustainable future, we must develop fuels that meet our needs for energy without sacrificing an equally important, though often overlooked, requirement: our need for a clean, healthful environment.

Resource supplies are also vital. From the long-term perspective, it makes sense to pursue those resources that are most abundant. And what could be more abundant than a renewable fuel supply?

In sum, when seeking alternatives to waning supplies of fossil fuels, we must

Fig. 1-2: *This solar hot water system on The Evergreen Institute's classroom building provides hot water for showers, washing, etc. It could be expanded to provide space heat as well.*

DAN CHIRAS

proceed with caution and intelligence. We need to develop energy resources that meet our needs and have the highest net energy yield, the most abundant supplies, and the lowest overall cost — socially, economically, and environmentally. In this book, I present up-to-date information on net energy yields to help you sort through the list of options. I'll also look at the pros and cons of various technologies, to help you make the wisest choices.

Before we go much further, though, let us take a brief look at energy itself. To help you understand this elusive entity, I'll cover some basics here, then introduce more concepts later in the book as we explore the various renewable energy systems.

UNDERSTANDING ENERGY

Like love, energy is all around us, but is sometimes difficult to define.

Energy Comes in Many Forms

In our quest to define energy, let's begin by making a simple observation: energy comes in many forms. For example, humans in many countries rely today on fossil fuels such as coal, oil, and natural gas to meet many of their energy needs. And some use nuclear energy derived from splitting atoms. In other countries, wood and other forms of biomass, like dried animal dung, are primary forms of energy. (Biomass includes a wide assortment of solid fuels, such as wood; and liquid fuels, such as ethanol derived from corn, and biodiesel made from vegetable oils; and gases such as methane, released from rotting garbage and animal waste.) And don't forget sunlight, wind, hydropower, and the geothermal energy produced in the Earth's interior. Even a cube of sugar contains energy. Touch a match to it, and it will burn — giving off heat and light, two additional forms of energy.

Energy Can Be Renewable or Nonrenewable

Energy in its various forms can be broadly classified as either renewable or nonrenewable. Renewable energy, as noted earlier, is any form of energy that's regenerated by natural forces. Wind, for instance, is a renewable form of energy. It is available to us year after year thanks in large part to the unequal heating of the Earth's surface. When one area is warmed by the sun, hot air is produced. This hot air rises and, as it does, cooler air moves in from neighboring areas. As the cool air moves in, it creates winds of varying intensities.

Renewable energy is everywhere and is replenished year after year. It could provide humankind with an enormous supply … if we're smart enough to tap into it.

Nonrenewable energy, on the other hand, is finite. It cannot be regenerated in a timely fashion by natural processes. Coal, oil, natural gas, tar sands, oil shale, and nuclear energy are all nonrenewable forms of energy. Ironically, most of these sources of energy are the products of natural biological and geological processes — processes that continue even today, but at rates not even remotely close to our rates of consumption. Coal, for instance, still forms in swamplands, but its regeneration takes place at such a painfully slow rate that it is impossible for the Earth to replenish the massive supplies that we are consuming at breakneck speed. Because of this, coal, oil, natural gas and the like are finite.

When they're gone, they're gone.

Energy Can Be Converted from One Form to Another

There's still more to this mysterious thing we call energy. Even the casual observer can tell you that energy can be converted from one form to another. Natural gas, for example, when burned, is converted to heat and light. Coal, oil, wood, biodiesel, and other fuels are also converted to other forms of energy during combustion. Heat and light are byproducts of these conversions. Heat, in turn, can be used to make electricity. But the possibilities don't end here. Visible light contained in sunlight can be converted to heat. It can also be converted to electrical energy by solar cells. Even wind can be converted to electricity or mechanical energy that can drive

a pump to draw water from the ground. Humans have invented numerous technologies to convert raw energy into useful forms.

Energy Conversions Allow Us to Put Energy to Good Use

Not only *can* energy be converted to other forms, it *must* be for us to derive benefit. Coal, by itself, is of little value to us. It's a sedimentary rock that looks cool, but it is the heat and electricity produced when coal is burned in power plants that are of value to us. Sunlight is pretty, and it feeds the plants we eat, but in our homes and factories, the heat the sun produces and the electricity we can generate from it are of great value to us.

In sum, then, it is not raw forms of energy that we need. It is the byproducts of energy conversion — new types of energy that are unleashed when we convert raw energy through the many ingenious energy-liberating technologies — that meet the many and complex needs of society.

Energy Can Neither Be Created nor Destroyed

Another important fact about energy is something you may have learned in high school, that is, that energy can neither be created nor can it be destroyed. Physicists refer to this as the *First Law of Thermodynamics* or, simply, the *First Law*.

The First Law says that all energy comes from pre-existing forms. Even though you may think you are creating energy when you burn a piece of firewood in a wood stove, all you are doing is unleashing energy contained in the wood — specifically, the energy locked in the chemical bonds in the molecules that make up wood. This energy, in turn, came from sunlight. And the sun's energy came from the fusion of hydrogen atoms in the sun's interior.

Energy is Degraded When it is Converted from One Form to Another

More important to us, however, is the *Second Law of Thermodynamics.* The Second Law says, quite simply, that when one form of energy is converted to another form — for example, when you burn natural gas to produce heat — it is degraded. Translated, that means energy conversions transform high-quality energy resources to low-quality energy. Natural gas, for instance, contains a huge amount of energy in a small volume; the energy is locked up in the chemical bonds that attach the carbon atom to the four hydrogen atoms of each methane molecule. When these bonds are broken, the stored chemical energy is released. Light and heat are the products. Both light and heat are less concentrated — and thus, lower quality — forms of energy. Hence, we say that natural gas, which is a concentrated form

of energy, is "degraded" when burned. In electric power plants, only about 50 percent of the energy contained in natural gas is converted to electrical energy. The rest is "lost" as heat that dissipates into the environment.

No Energy Conversion is 100 Percent Efficient (Not Even Close)

This leads us to another important fact about energy: no energy conversion is 100 percent efficient. When coal is burned in an electric power plant, only about 30 percent of the energy contained in the coal is converted to useful energy — in this case, electricity. The rest is lost as heat and light. The same goes for renewable energy technologies. One hundred units of solar energy beaming down on a solar electric module won't produce the equivalent of 100 units of electricity. You'll only get around 8–20 percent conversion using the various solar modules on the market today. (However, some new technologies can capture and convert about 35 to 40 percent of the incoming energy. So, things are improving.)

Energy is lost in all conversions. As another example, most conventional incandescent lightbulbs convert only about five percent of the electrical energy that runs through them into light. The rest is released as heat. (Incandescent lightbulbs should really be called "heat bulbs.")

Each conversion in a chain of energy conversions loses useful energy, as shown in Figure 1-3. To get the most out of our primary energy sources, therefore, we have to reduce the number of conversions along the path.

You may be wondering if all of this discussion of energy losses is a violation of the First Law of Thermodynamics, which states that energy cannot be created or destroyed. The answer is no. The energy losses that take place during energy conversion are not really losses in the true sense of the word. Energy is not destroyed; it is released in various forms, some are useful to us and others, such as heat, are not so useful. Chemical energy in gasoline propels a car forward along the highway. Some is also lost as heat that radiates off the engine. This waste heat is of little value — except on cold winter days when it is used to warm the car's interior. Eventually, however, all the heat produced by a motorized vehicle escapes into outer space. It is not destroyed, per se, but it is no longer available to us. Hence, the conversion results in a net loss of *useful* energy.

OK, so now you're "cooking" with information about energy. You know there are many forms of energy. You know that

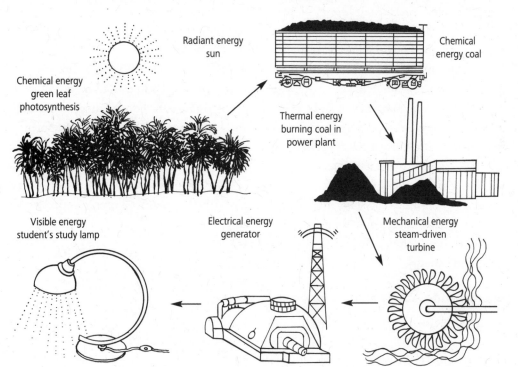

Radiant energy
sun

Chemical energy
coal

Chemical energy
green leaf
photosynthesis

Thermal energy
burning coal in
power plant

Visible energy
student's study lamp

Electrical energy
generator

Mechanical energy
steam-driven
turbine

Fig. 1-3:
Energy conversions occur commonly in the production-consumption cycle of various fuels. Unfortunately, none of these conversions is 100 percent efficient, so energy is lost at each stage. The key to using energy efficiently is to limit or eliminate conversions.

Energy, quite simply, is valuable because it allows us to perform work. It powers our bodies. It powers our homes. It powers our cars and trucks. We cannot exist without energy.

energy can be renewable or nonrenewable. You understand that raw energy is not as important to us as are its useful byproducts such as electricity, light, or heat. You also know that energy can be neither created nor destroyed; it can only be converted from one form to another. Raw energy is useful to us because it can be converted into other forms. And you're privy to the fact that no energy conversion is 100 percent efficient, not even close.

You also understand that during conversions, useful energy decreases. That fact, in turn, is important for nonrenewable fuels; once they've burned, or reacted, in the case of nuclear fuels, their energy is gone forever. The heat radiates endlessly into outer space, heating the universe, as it were. Renewable energy resources, on the other hand, can be regenerated year after year after year. If we're going to persist as a society, it is renewable energy resources we'll need to rely on. Unlike fossil fuel energy and nuclear energy, renewable resources can be regenerated continuously as long as the sun continues to shine, making our lives bright and cheery and comfortable.

With these important points in mind, let's define energy.

Energy is the Ability to Do Work

To a physicist, energy is "the ability to do work." Any time you lift an object, or slide an object across the floor, you are

performing work. The same holds for our machines. Any time a machine lifts something or moves it from one place to another, it performs work.

According to physicists, work is also performed when the temperature of a substance is raised. Therefore, your stove or microwave is doing work when it boils water for hot tea or soup.

Work and Power

Physicists aren't content to just define energy; they also like to measure it. Because the units of measurement they use come in handy when sizing renewable energy systems, it is worth learning a few of the most common ones.

Measuring Mechanical Energy

If you push a 100-pound box of potatoes 10 feet across a wooden floor, physicists would proclaim that you have performed work. And you'd probably agree. But how much work have you performed?

To calculate the amount of work performed, scientists use a very simple equation:

$$\textbf{Work} = \textbf{Force x Distance}$$
In this case,
work = 100 pounds x 10 feet
= 1,000 foot-pounds of work.

Although "foot-pounds" isn't a particularly useful term, the example shows you what is meant when we say that energy allows us to do work.

It is also important to be able to measure the rate at which work is performed. Physicists define the rate at which you or a machine performs work as power. An adult, for instance, might be able to push a 100-pound box 10 feet across a floor in 4 seconds. It might take a 10-year-old 10 seconds to perform that same amount of work. Clearly, the adult is more powerful. Even though they perform the same amount of work, the adult performs it more quickly.

For the mathematically inclined: power is the force x the distance (or work performed) divided by the time it took to complete the task. Another way of writing this is:

Power = Work/Time.
In the case of the adult,
power = 1,000 foot-pounds/4 seconds,
or 250 foot-pounds per second.

In the case of the child,
power = 1,000 foot-pounds/10 seconds,
or 100 foot-pounds per second.

I bring this up not to torment you with physics, but because you will encounter units of work and power as you study renewable energy options. However, in this book, most of our attention will focus on thermal and electrical energy.

Thermal, or heat, energy can be measured in BTUs (British Thermal Units).

One BTU is the amount of energy it takes to raise one pound of water one degree Fahrenheit. Remember, as just noted, raising temperature is a measure of work.

Furnaces, boilers, and water heaters are all rated by their BTU output. Solar water heaters are rated — and compared to one another — by the number of BTUs of heat they will generate under controlled conditions (so you can see how one stacks up against another). In addition, energy auditors, whom you may be hiring, typically calculate heat loss from buildings in BTUs — usually BTUs per square foot per year. Solar heating specialists calculate heating loads (how much heat a home needs) in BTUs as well. Cooling requires a calculation of how much heat in BTUs must be removed from a building per hour to maintain comfort.

A single BTU is not much energy when compared to what it takes to provide hot water for a family of four or to heat a home for a year. Most of the time, you will be dealing with tens of thousands of BTUs or, in the case of home heating, millions. A house, for example, can gain tens of millions of BTUs of heat energy from the sun each year.

While the BTU is a measurement of heat energy, electrical energy has its own set of terms that you'll need to be familiar with if you decide to do something about your home's electrical energy consumption.

While I've got you on the torture rack, let's take a look at them now. The pain will ease up shortly.

Electricity is really the flow of electrons, tiny subatomic particles, through wires. To help students understand electricity, most teachers liken electricity to water flowing in a hose; the water molecules are akin to the electrons flowing in a wire.

As you know, water can flow slowly or very quickly. How fast water flows is determined by a force: water pressure. Water flowing out of the hose of a rain barrel, for example, flows pretty slowly. Water flowing through a downspout from the gutter that drains a roof flows more quickly. It is under more pressure than the water flowing out of the rain barrel.

How fast electrons flow through a wire also depends on a force much like the water pressure in a hose. This force is measured in volts. Voltage is an *electromotive force* — a force that causes electrons to move.

In the world of electricity, electrical power can be calculated using the same equation we used to determine mechanical energy:

Power = Force x Distance/Time.

So, in electrical energy, the force is the volts. What about the distance and time components of the equation?

Electrons can flow at different rates in wires, too. The number of electrons flowing past a point in the wire at any one time is measured as amps, short for amperes. Multiplying volts (force) times amps (distance/time), therefore, gives you a measure of power. Power is measured in a term familiar to most of us: watts. To simplify matters, physicists calculate watts using the following equation:

Watts = amps x volts

Watts is a common term that you need to become familiar with. Fortunately, most of us already are familiar with watts. When we shop for a new microwave, we shop by watts: This unit is a 1,000-watt microwave; that one is a 1,200-watt unit. The higher the wattage, the faster it will cook our food. We've all shopped for 100-watt lightbulbs or new compact fluorescent lightbulbs that produce the same as a 100- or 75-watt incandescent.

Watts is used as a measure of instantaneous power *consumption* of motors, electric hair dryers, and lightbulbs. Watts is also used to measure power *production*. A 4,000-watt solar electric system, for instance, produces 4,000 watts on a bright sunny day. A 10,000-watt wind turbine produces 10,000 watts when the wind is blowing at a certain speed (usually around 26 to 30 miles per hour). Since 1,000 is a kilo, a 1,000-watt solar electric system is a 1-kilowatt system or 1-kW system. A 10,000-watt wind turbine is a 10-kW wind turbine.

Although wattage is a measure of instantaneous power production, it's not the instantaneous production of power that matters to us when sizing home energy systems, but power production *over time* — measured in watt-hours and kilowatt-hours. A 1-watt lightbulb shining for one hour consumes 1 watt-hour of energy. A 100-watt lightbulb burning for 10 hours consumes 1,000 watt-hours of energy, or 1 kilowatt-hour.

Utility companies bill our energy use measured in kilowatt-hours, and it is the kilowatt-hour production of solar and wind systems that are of greatest concern to those who are thinking about turning to solar or wind to meet their needs for electricity.

Ok, let's stop here and turn our attention to the shining star of the energy show, renewable energy.

WHAT IS RENEWABLE ENERGY?

Renewable energy, as just noted, is a form of energy capable of being regenerated by natural processes at meaningful rates. Wood, for instance, is a form of renewable energy. It's produced by trees from minerals in the soil, carbon dioxide in the air, water from the ground, and energy from the sun. Wind, flowing water, and sunlight are also

DAN CHIRAS

renewable energy resources (Figure 1-4). Heat within the Earth's crust can be considered to be renewable, too, as are the tides.

Most of these renewable forms of energy are made possible by the kingpin of all renewables, the sun, the center star of our solar system.

Let's get something out of the way right from the start, however. The sun is really *not* a renewable energy source. This blisteringly hot ball of gases is a huge fusion reactor, near the center of our solar system, 93 million miles from planet Earth. Fusion reactions occur between hydrogen atoms in the sun's interior. When two hydrogen atoms fuse, they form a helium atom and enormous amounts of heat, light, and other forms of energy. That's what makes the sun tick.

Fig. 1-4:
The energy of falling water can be captured on a small scale or a very large scale. Although this form of energy is relatively clean, damming rivers can create enormous environmental impacts.

And the Earth, too.

Although only about 0.5 billionths of the sun's energy strikes the Earth, that is enough to power virtually all life on land and sea, and it has been responsible for the build-up of vast resources of fossil fuels that are now quickly (in geological and human time) sliding toward oblivion.

Technically, though, the sun is a finite resource. Someday its fuel source will run out. But don't fret; there's good news: the sun is going to burn brightly for at least five billion years or more before it dies out on us. And when it does, it will be all over. "We don't have five billion years of sunshine, however. Scientists calculate that the sun's energy output will slowly increase over the next billion years, and in about one billion years, the sun will be so hot that it will extinguish all life on planet Earth."

Although the sun is a finite resource, teachers, energy experts, and books on the subject still refer to it as a renewable energy resource.

THE PROS AND CONS OF RENEWABLE ENERGY

Before examining the pros and cons of renewable energy, it is important — indeed vital — to point out a dangerous trap that many fall into, notably, lumping renewables into a single category. Renewable energy encompasses a half dozen or so fuels — solar, wind, hydropower, biomass, hydrogen, and geothermal — and a number of different technologies.

Because there are many renewable energy technologies, it is misleading to lump them into broad categories. It is inaccurate, for example, to speak of solar energy technologies as a single entity; granted, they are fueled by a common fuel source — the sun — but there are a number of very different technologies that derive their energy from the sun: solar electric cells, passive solar, solar hot water collectors, and solar hot air collectors. (See Figure 1-5.) And each of these technologies has different uses, benefits, challenges, and costs. Troubles often arise when critics of renewable energy lump all available technologies together. For example, opponents of solar energy are fond of saying that "solar energy is too expensive," which is a little like saying, "All Americans are idiots," just because there are a few nitwits amongst us.

Although it is true that some solar technologies like solar electricity are expensive (at least currently), others are quite cost-competitive with conventional technologies. A good example of the latter is passive solar heating. That said, there are times when the most expensive technologies like solar electricity or photovoltaics produce electricity at a lower cost than conventional sources. In less developed countries, for instance, it's often cheaper to install solar electric modules in rural villages than to string power lines hundreds of miles from existing coal-fired power plants to service such remote locations.

Even in industrialized nations, the cost issue is not always straightforward. Solar electricity can be less expensive than conventional electrical power in some locations. For example, when powering highway warning signs, it is much cheaper to install a solar electric-powered sign than to run an electric wire to the unit (Figure 1-6). Emergency call boxes are similarly powered for the same reason (Figure 1-7). Moreover, if you build a home just two- or three-tenths of a mile from a power line, you'll find that it is often less expensive to install a full-scale solar electric system than it would be to string electric wires from the utility's power line to your home. In fact, running an electrical line a couple of tenths of a mile might cost $20,000 to $30,000. And that doesn't give you a single kilowatt-hour of electricity. All it provides is access to the utility's grid.

Financial incentives make solar electricity much more cost-competitive to homeowners. Given various local, state, and national incentives, at least in the United States, homeowners can install a pretty impressive solar electric system for $30,000. And if you build more than half a mile from an existing power line, a solar

Fig. 1-5: *These solar cells (a) are made of silicon, which comes from silicon dioxide from sand and quartz. Solar cells convert solar energy into electricity. Many solar cells are wired together in a module which is mounted on the roof or on a free-standing rack or pole (b) like these arrays at the Solar Living Institute in California. The array on the right side of the photo is mounted on a tracker that points the array toward the sun from sunrise to sunset.*

Fig. 1-6:
This flashing stop sign in rural Missouri warns drivers of a very dangerous intersection. It is powered by solar electricity.

Fig. 1-7:
Emergency phones along a remote northern California highway are made possible by solar electricity. Stringing electrical lines to such locations would be cost-prohibitive.

DAN CHIRAS

system is hands-down, without question, much less expensive than connecting to the electrical grid — unless your utility picks up the tab for running the wire to your home. Further clouding the cost issue, some states or utilities in states such as California, Wisconsin, Illinois, New York, New Jersey, Colorado, and Missouri offer generous incentives to homeowners and business owners that reduce the cost of solar electric systems. State, local, and federal incentives can reduce the initial cost for homeowners by up to 50 percent. Businesses may receive even greater benefits for solar electricity, reducing the cost of a system by about two thirds.

Solar electricity also makes sense in cases where grid power is expensive. In California, for instance, electricity from the grid during peak usage periods typically runs from 30 cents to over a dollar a kilowatt-hour, far more than the cost of generating electricity from sunlight. Grid power in Germany is currently 75 cents per kilowatt-hour, creating a bonanza for US solar cell manufacturers.

When listing the pros and cons of renewable energy technologies, then, we need to be very specific. Keep this in mind when discussing renewable energy technologies with friends, government officials, or opponents. Watch out for people who unfairly lump renewable energy technologies into one group.

With this friendly advice in mind, we can now look at some of renewable energy's pros and cons.

The Benefits

Renewable energy offers many benefits, outlined here.

Renewable energy resources are available year after year — often in predictable amounts in all parts of the world. Thus, they're fairly reliable resources — energy sources we can count on — and they're renewable. They're not finite like coal, oil, natural gas, and nuclear.

Another big advantage of renewable energy technologies is that, with few exceptions, the fuel is *free*. It's not under the control of oil cartels or wealthy multinational corporations. Although it costs money to tap into these renewable energy resources, no one controls the price of most renewable fuels, so we won't experience rising prices. Installing a renewable energy system, therefore, serves as a vital protection against rising fuel prices experienced as finite resources like oil and natural gas dwindle. Renewables will provide a hedge against inflation!

Renewable energy is *clean and non-polluting;* except for the manufacture and installation of the technologies needed to capture these forms of energy, there's very little, if any, pollution or environmental damage created by the use of most renewable energy technologies. Renewable energy technologies are the cleanest form of energy, next to energy conservation and energy efficiency measures.

For the most part, the technologies required to convert energy from renewable sources to useful energy are *already available*. No new breakthroughs are required for their widespread use, although improvements in technology and new developments in some areas could help lower their costs and make some sources more practical.

Many renewable energy technologies and fuels are *cost-competitive* with conventional technologies and fuels. Right now, large-scale wind power, passive solar heating, passive cooling, solar hot water, solar hot air, solar electric systems in rural settings or in regions with high electric costs are frequently cost-competitive alternatives to fossil fuels and nuclear energy.

Renewable energy technologies are, for the most part, *decentralized sources of power*, not vulnerable to sabotage. And, as just noted, they're not controlled by multinational corporations. Much of the renewable energy business currently consists of smaller companies, with a few big dogs like General Electric and British Petroleum starting to play key roles.

Renewable energy technologies permit individuals to *control their own power production*. When you install a solar electric system or a wind generator, you become your own plant manager. You're no longer at the mercy of your local power company. As a result, renewable energy helps us attain personal freedom and independence.

For those who love gadgets, *renewables are fun*. There's hardly a day that goes by that I'm not reading the meters on my solar electric, solar hot water, or wind systems and making graphs to track how much energy I'm getting from each one.

The Disadvantages

As with all things in our lives, there are downsides to renewable energy, many of which we can easily work around or eliminate with a little ingenuity.

Renewable energy technologies and renewable energy fuels constitute only a small portion of our current energy production and consumption, so *they have fewer advocates to promote their use*, although that is starting to change through the efforts of progressive governors and legislators and big companies such as General Electric and Vestas.

Renewable energy is often *actively lobbied against* by the powerful fossil fuel industry, for example, coal companies or the nuclear industry. They and others often spread misinformation about renewable energy to make their case to the public. With such powerful forces working against renewable energy, creating a renewable energy future won't be easy.

Compared to conventional fossil fuels, renewable energy research and development is *poorly funded in most countries*, the United States and Canada being prime examples. Here again, there are notable exceptions. In Europe, several countries such as Germany, Great Britain, and Holland are making tremendous efforts to tap into renewable energy. In recent years, the United States has taken a much more active role in promoting renewable energy as part of the nation's clean energy policy, especially under the leadership of Barack Obama. Generous tax incentives and training programs have helped spur the industry.

Renewable energy resources such as wind and solar are *not available 24 hours a day, 7 days a week, unlike fossil fuels*. Renewable energy technologies often require some means of storage so surpluses can be stockpiled for later use. Storing renewable energy can be difficult and costly. That said, solar and wind system owners meet this challenge by installing batteries that store surplus during times of excess and provide electricity when the sun or winds are not available. Serious efforts are underway to develop long-term, massive energy storage systems for large commercial systems.

Conventional fuels may help meet our needs when the sun or wind are not available. In the future, for example, the world's renewable energy system will very likely consist of a combination of renewable technologies — such as wind, hydropower, and solar — with backup energy from nonrenewable technologies, such as natural gas and even new coal-fired power plants. Facilities like natural gas and new coal-fired power plants can be brought up online quickly, providing a steady supply of energy to meet our needs.

Regional energy transfer may also help ensure a continuous supply of energy. Deficiencies in one part of a country, for instance, can be offset by surpluses generated in others, as is common in the electric-generating industry today. Electricity generated by wind farms in Kansas, for example, could be fed to Colorado or New Mexico to meet their needs during cloudy or windless periods.

Even though many renewable energy technologies are not available 24 hours a day, seven days a week, the amount available in a given year is quite predictable. That is, you can count on a certain amount of sunlight and wind. Individuals can design systems that generate surplus during times of excess for use in times of shortage. Net metering policies in some states make this quite simple.

Another downside of renewable energy is that there are *not enough experts and local suppliers of renewable energy technologies.* Open the Yellow Pages and check listings on solar hot water installers and compare the number of these companies to the number of companies that install conventional water heaters. In the vast majority of the country, the latter list is much longer.

Some technologies are quite *expensive and not yet competitive with conventional fuels* at current prices, although this could change with rising oil and natural gas prices and financial incentives. That said, as mentioned earlier, many renewable energy technologies are quite affordable — even cheaper than conventional technologies.

Converting to renewable energy technologies may need to be done on a house-by-house basis, which will require the efforts of millions of informed homeowners.

With this list of pros and cons in mind, let's begin our exploration of the various renewable energy technologies that are available to homeowners and business owners. Along the way, you'll learn more about the pros and cons of each energy source/technology. I also hope to shatter a few myths that are holding back this important revolution in home and business energy production.

PROSPECTS FOR THE FUTURE

For 14 years I ran my entire Evergreen, Colorado house and office almost 100 percent off renewable energy — sunlight, wind, and firewood. I used only a small amount of natural gas for heating water

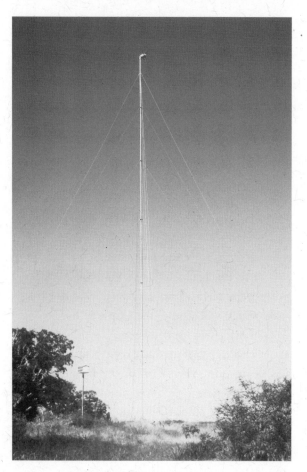

Fig. 1-8: *On top of this 126-foot tilt-up tower, we fly a 2.5-kW wind turbine that provides a good portion of The Evergreen Institute's electricity.*

and cooking meals. I currently run The Evergreen Institute's Center for Renewable Energy and Green Building in Missouri *entirely* off solar and wind energy (Figure 1-8).

You can live independently, too.

As you will see throughout this book, energy independence is almost always more easily attained by those building anew than by those retrofitting a home

for self-sufficiency. When building a new home, simply orienting it to the south to increase winter heat gain and reduce summer heat gain can cut one's heating and cooling bills by 10 percent. By concentrating windows on the south side of the home, a homeowner can cut heating and cooling bills by up to 30 percent. But like many readers, you probably don't have the luxury of starting from scratch. You have a home that's firmly anchored to the ground and, most likely, not ideally situated for optimal solar gain. If you want to retrofit your home for greater energy self-sufficiency, you'll face a much bigger challenge than the lucky person who builds anew.

Don't be discouraged, however.

You can easily cut your energy demand by half, perhaps even more, with some simple, cost-effective measures that improve the energy efficiency of your home. A few more "aggressive" energy efficiency and conservation steps could allow you to slash your fuel bill even more.

With these measures in place, you can install some renewable energy technologies that will bring your household closer to full energy independence. In doing so, you'll not only reduce your current monthly fuel bill, you'll help protect yourself from the potentially devastating rise in fuel costs. You'll also help reduce greenhouse warming and a host of other serious environmental problems, and you make it easier for the United States and Canada — and all other nations — to meet their energy needs in a sustainable manner. The more of us that take these steps, the better our nations' futures. This stuff is vital to national security.

Fortunately, there are ample supplies of renewable energy in many areas. Many renewable energy technologies such as solar hot water and residential-scale solar electricity are easy to install in a business or residence. Others, like large-scale wind power, will require the deep pockets of private industry and the far-sighted assistance of local, state, and federal governments.

In our quest for a better, brighter energy future, we should never lose sight of two facts. First, there's much we can do to improve energy efficiency in our homes. A huge amount. Moreover, we can improve energy efficiency without sacrificing services we've become accustomed to. We can live lightly and live well! Conservation isn't "freezing in the dark," as former President Ronald Reagan was fond of saying. Conservation is living comfortably at a fraction of the cost of our wasteful lifestyles. Conservation, if we're smart, means a better life for us.

Second, renewable energy resources are vast. They outshine the remaining nonrenewable energy resources by so much, it makes your head spin. Enough sunlight strikes the Earth in 40 minutes to power *all* of our energy needs for a full year!

You may be surprised to learn that renewable energy, even sunlight, can be used to power nearly all of your family's needs — even in some of the gloomiest parts of the United States. Renewable energy and conservation, then, are the key to a sustainable future, and you and millions of people like you can turn that key.

I encourage you to think outside the energy box, too. Look for avenues that lead to energy *savings*, for example, growing more of your own food. Home gardens can save huge amounts of energy, and can help create a more independent and sustainable lifestyle. But what if you don't have room in your backyard, or don't even have a backyard? You can be part of a community garden. Community gardens are often placed in vacant lots in cities and towns. They allow people to grow much of their own food. By growing your own food, you help reduce the need for the massive amounts of energy currently used to produce and ship food from farms all over the world to people like yourself.

Travel consumes huge amounts of energy, too, as does the production of clothing and other goods we buy. Buy locally. Consider vacationing locally. Consider used goods. Recycle everything you can. Buy recycled materials whenever possible. Thinking beyond heating, cooling, and lighting will help you place yourself on the path to energy independence and a cleaner, healthier, safer future. Renewable energy and conservation are the key to a sustainable future. You, and millions of people like you, hold that key.

CHAPTER 2

CONSERVATION RULES

THE CORNERSTONE OF YOUR ENERGY FUTURE

A few years after I moved into my super-efficient passive solar/solar electric home, one of my neighbors said, "I really admire what you do, Dan, but I couldn't live like that." I was puzzled by her statement, and stood there speechless for a while, wondering what to say. I finally offered up a feeble response: "We have all of the conveniences of a modern home," I said, stretching the truth a bit. We've passed up opportunities to buy or receive (as gifts) various modern "conveniences," such as an electric can opener or electric carving knife. (The mechanical, non-electric options work just as well — or even better, in my opinion — and require fewer materials to make and less energy to operate.)

As I've traveled throughout North America giving talks on renewable energy and green building, many other individuals I've run across have reacted similarly to the apparently odd notion of living on home-based renewable energy. Men sometimes say, "I like the idea of solar electricity, but I couldn't do it because I hear power tools won't work."

The fact is, you can live any way you want using solar electricity — or any other modern renewable energy technology. You can have a big screen TV, or two. You can run your lights 24 hours a day and use any power tool you'd like. If you choose to live this extremely energy-intensive (and wasteful) lifestyle, however, be forewarned: it will likely cost you an arm and a leg to supply your needs with renewable energy!

Consider this true story: Many years ago I lived in a passive solar home that had

31

been designed by a local, energy-conscious solar builder (Figure 2-1). After reading a book on solar electricity, I decided I wanted to install a solar electric system on the house. Unfortunately, this marvelous house — while designed to be heated passively — came with an electric stove, an appliance that eats electricity for breakfast, lunch, dinner, and snacks. As if that weren't bad enough, the house also came with an electric backup heating system. It too had a voracious appetite for electricity.

Undaunted, I ran the calculations to determine exactly how much it would

cost me to install a solar electric system to generate electricity for this otherwise energy-efficient home. When finished, I was aghast: by my estimations, the solar electric system would cost a whopping $50,000 — in 1987 dollars! Needless to say, I never installed a system on that home. Even with all of the energy-saving strategies in place, the cost was just too high.

Years later, I designed and built a home of equal size. Unlike the contractor who had built my previous home, I paid very close attention to energy efficiency at every step in the process. I installed energy-efficient lighting. I designed the house for daylighting — a technique that provides as much natural lighting as possible during daytime hours. I installed a super-efficient refrigerator and a top-of-the-line energy-efficient washing machine. I even purchased energy-efficient televisions. Rather than installing an electric stove, I put in a gas stove. And instead of installing an electric baseboard heating system for backup, I installed an energy-efficient radiant floor heating system (I later replaced it with a wood stove because there's so much free firewood in my area. Someone's always thinning the trees on their property for fire safety or clearing a lot on which to build, so there are mountains of free firewood for the asking.)

Through careful attention to detail, I was able to slash my projected electrical

DAN CHIRAS

Fig. 2-1: *My first passive solar home, while heated by the sun, relied heavily on electricity for backup heating and cooking. The demand was so high, despite energy conservation measures, it would have cost me around $50,000 to install a solar electric system. There were no incentives at the time, so I decided to pass on solar electric and took numerous steps to reduce energy demand through efficiency.*

energy demands by 75 percent. I was also able to minimize heating demands, cutting them by 80 to 90 percent through a combination of renewable energy (notably, passive solar design) and a host of energy-efficiency measures, among them super-insulation, airtight construction, and earth sheltering.

As a result of these and a host of other design ideas, I was able to dramatically reduce the size of the solar electric system. Instead of $50,000, my system cost around $15,000. Although it is a very small system, it supplied *all* of our electrical energy needs (with a little help from a wind generator) for the 14 years that we lived in the house. The system allowed my children and me to enjoy virtually all of the amenities of modern life, including two televisions, a stereo, a microwave oven, a blender, a computer, power tools, and others. We even had a bathroom ceiling fan!

We didn't leave lights on day and night like my neighbors, who worried about being able to live the way I did. But the bottom line was: we lived well using only a fraction of the energy most households use. You can, too.

However, as this story suggests, the cornerstone of any sensible plan to switch to renewable energy is first and foremost to slash energy use through conservation and efficiency.

EFFICIENCY FIRST!

In this book, I use the term *energy conservation* to refer to simple measures to cut energy demand — for example, wearing a sweater on cold winter nights or shutting off lights and electronic equipment when not in use. Most conservation strategies are simple and painless. All of them offer amazing return on your efforts. You already know the items on this list: turn off lights, televisions, and stereos in unoccupied rooms, take shorter showers to reduce hot water use, etc.

I use the term *energy efficiency* to refer to measures to wring more useful energy — BTUs and watts — from technologies that make our lives more comfortable and convenient. Energy-efficient compact fluorescent lightbulbs (CFLs), for instance, convert 25 percent of the electricity flowing through them into light (Figure 2-2). A standard incandescent lightbulb converts only 5 percent of the electricity flowing through it into light. Compact fluorescents are the hands-down favorites when it comes to energy efficiency. Even better are the new LED lights (LED stands for *light-emitting diodes*). They use much less energy than CFLs and very likely will replace CFLs in the next two decades (Figure 2-3).

Consider another example: water-efficient showerheads. These simple devices use one half to one third as much water as

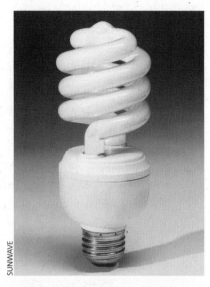

Fig. 2-2:
Compact fluorescent lightbulbs like this one use one fourth of the energy of standard incandescent lightbulbs, produce much less heat, save big on energy bills, and dramatically reduce carbon dioxide emissions by using less electricity.

SUNWAVE

DAN CHIRAS

Fig. 2-3: *This energy-efficient floodlight installed at The Evergreen Institute faculty residence uses only 10 watts and will last approximately 50,000 to 60,000 hours.*

standard showerheads. Less hot water means less energy. Much less energy.

Both the compact fluorescent lightbulb and the efficient showerhead perform their duties quite admirably while consuming only a fraction of the energy of their less energy-efficient counterparts that are found in many homes today. Bottom line: by switching to energy-efficient devices, you'll receive the same service — or even better service — at a fraction of the cost. Combined, conservation and efficiency can result in huge savings, making them indispensable components of our home (and office) energy strategies. To understand the economic benefit of such measures, consider first the impact of efficiency alone. Let's suppose that you have a security light on your front porch fitted with a 100-watt incandescent lightbulb. The light operates 24 hours per day, 365 days a year because you never remember to turn it off when leaving for work. If you replaced that 100-watt incandescent lightbulb with an energy-efficient compact fluorescent lightbulb, what would it save you?

A 100-watt incandescent lightbulb left on 24 hours a day, 365 days a year will operate 8,760 hours each year. To determine how much energy the bulb consumes, simply multiply the wattage of the bulb (its power demand) by the number of hours it operates. When you do, you find that the bulb consumes 876,000 watt-hours of electricity per year — that's 876 kilowatt-hours (kWh) per year. (Remember from Chapter 1 that 1,000 watt-hours is a kilowatt-hour [kWh], which is the unit of measurement utilities use to charge you for electricity.)

The compact fluorescent lightbulb you replaced it with uses one fourth of the electricity. By changing the bulb, your monthly electrical use plummets from 876 to 219 kWh per year. If you pay 10 cents per kWh of electricity, your annual bill for this single bulb will drop from $87.60 to $21.90, a savings of $65.70 per year. Efficiency clearly pays!

Now, let's suppose that you also turn the compact fluorescent lightbulb off each morning on your way out the door (a conservation measure); now the porch light only operates 12 hours a day. The cost of running the lightbulb is now just $11.00 per year, another $10.90 in savings. By combining efficiency (use of the more efficient bulb) with conservation (turning the light off during the day), you'd reduce electrical consumption to around 110 kilowatt-hours per year. The grand total of your savings from frugality and efficiency would be $76.60 per year. And that's just for a single porch light!

But that's not all. To compare the economic cost of your current energy use plan to the proposed energy efficiency/ conservation plan, you also need to factor in the costs of the lightbulbs themselves. To do this, you have to determine how long each bulb lasts. Standard 100-watt incandescent lightbulbs sold at my local hardware store are rated at 860 hours each. That means that each bulb in a packet will last, on average, about 860 hours. A typical compact fluorescent for sale on the same aisle, however, is rated at 10,000 hours. How does this affect the cost?

A compact fluorescent installed in your porch light and left on 12 hours per day will last a little over 2 years (2.2 years to be exact). An incandescent lightbulb, which has a much shorter lifespan, will burn out approximately every two and a half months.

Over the lifespan of one CFL (2.2 years in this situation), then, you'd need to purchase and install 10 incandescent lightbulbs. Compact fluorescent lightbulbs cost a little more, typically about $1–$2, if you buy them in packs of three to six bulbs. Incandescent lightbulbs cost about 75 cents each, give or take a little. In this example, you would need 10 bulbs to equal one compact fluorescent lightbulb, and the total cost of the incandescent lightbulbs would be $7.50. The compact fluorescent lightbulb would cost $2.

So, in that two-year period, incandescent lightbulbs would cost you $175.20 in electricity and $7.50 for bulbs, for a

Table 2.1 Energy Savings from Conservation Per Year			
Type of bulb	100 watt incandescent light bulb	25 watt compact fluorescent	25 watt compact fluorescent
Hours on per day	24	24	12
Electrical consumption	876 kWh	219 kWh	110 kWh
Cost at 10 cents per kWh	$87.60	$21.90	$11.00
Annual Savings in electrical bill	--	$65.70	$76.60

total of $182.70. The CFLs would cost you $43.80 for electricity and $2 for the lightbulb, for a total of $45.80. This strategy would save you $136.90. The CFL strategy (efficiency) combined with conservation (shutting off the light each morning) would cost $22 for electricity and $2 for the bulb, bringing the total to $24. This strategy would save you $158.70. Either strategy represents a huge return on your investment. If only our retirement accounts performed one tenth as well!

As this example shows, although installing an energy-efficient compact fluorescent lightbulb may cost a little more up front, it will save you a huge amount of money in the long run.

As you think about reshaping your energy future, bear in mind that you can —

and should — tap into both efficiency *and* conservation measures. With a little exercise of your gray matter, you can find dozens and dozens of ways to reduce energy demand by being more energy conscious and more efficient. Try to enlist the rest of your family, too. Pitching in together, your family can make tremendous strides in reducing overall energy use in your home.

Let me point out, however, quite emphatically, that energy conservation and energy efficiency don't require us to live an austere life devoid of pleasures. They simply mean making decisions that eliminate massive amounts of waste in our homes, places of business, and most of our lifestyles. It means living well — staying warm in the winter and cool in the summer. In fact, many measures can improve the quality of our lives. Sealing up the many leaks in the walls of a home, for instance, makes our homes much more comfortable throughout the year, while cutting energy demand. Energy conservation means saving money, enormous sums of money, that families can use for a host of other things, for example, retirement funds, vacations, or financing the children's college educations. (I'm pretty sure that most readers can find better ways to spend their hard-earned money than paying their utility companies!) These measures also reduce our demand for natural resources and the enormous environmental damage that results from the exploration, extraction, transport, and combustion of fossil fuels.

In sum, energy conservation and energy efficiency can eliminate discomfort, slash fuel bills, save money, and help create a cleaner, more sustainable world — a brighter future for ourselves, our children, their children, and the millions of species that share this planet with us.

In this chapter, we'll explore concrete ways you and your family can put a lid on energy consumption in your home (or office or cabin) while dramatically improving comfort. As you will soon see, saving energy can save you and your family tens, if not, hundreds of thousands of dollars over your lifetime.

It's no exaggeration. I've run the math. Using energy efficiently is like giving yourself a big fat pay raise. You should do it, even if you are not contemplating installing renewable energy technologies. Don't wait for your boss to open his checkbook. Go ahead and do it yourself!

If you are seriously thinking about installing renewable energy technologies on your home, even the most cost-effective options available today, energy efficiency and conservation should be your first steps. It will make the task easier and less expensive. When you do install a system in an energy-efficient home or business, you need a smaller one, which will be easier to install in your home or business. Smaller systems cost less. Why spend $40,000 for a solar electric system when a three or four thousand dollar investment in energy efficiency could enable you to install a much smaller system costing, say, $20,000–$25,000? Efficiency and conservation, therefore, pay huge dividends in the long run and will dramatically reduce the initial capital outlay for a renewable energy system.

Richard Perez, founder of *Home Power* magazine, estimates that every dollar you spend on efficiency will save you $3 to $5 on system costs. An investment of $2,000 in air sealing and better insulation and a more efficient refrigerator will reduce the initial cost of the solar electric system you need by $6,000 to $10,000 (before incentives). Because they are free, energy conservation measures offer even greater benefits — so get in the habit of managing your demand now, and you will save enormously on your renewable energy system.

ENERGY CONSERVATION AS A RENEWABLE SOURCE OF ENERGY

In most discussions on energy, energy conservation and efficiency are highly praised but mischaracterized. Most see it as a way of simply saving energy. To me, however, energy conservation and efficiency represent a way of *producing* energy — and producing it renewably.

Let's suppose that you begin by shoring up the leaks in the building envelope of your home by caulking the many tiny openings where pipes or wires penetrate exterior walls or where there are tiny gaps around doors and windows, or where the foundation and walls meet. You add insulation to the attic of your home, boosting its R-value (the measure of resistance to heat movement) to R-50. Next, you install storm windows to prevent heat loss from leaky single-pane windows. All of these activities cut down on your family's annual fuel bill and save money — lots of it.

The energy you save, however, can be used by others — say, the occupants of a brand new home across town. Your savings, therefore, become their energy source. Although you haven't actually created new energy, the net effect is just the same. Why?

Let's suppose you had watched a baseball game on TV or gone shopping instead of investing time and energy into making your home more energy-miserly. Had you lazed around, the local utility company would have had to procure more energy from some other source to meet the demands placed on their system by their new customer. In other words, they would have had to procure more coal or natural gas to meet the newcomer's electrical and heating needs. By simply cutting back on your use, you've saved them the work. You supplied the energy they needed.

The energy conservation sources you've tapped into are also renewable. The caulk applied to seal cracks in the building envelope of your home and the weather stripping used to seal doors and windows save energy that is available to others. And it keeps on yielding its savings year after year.

Insulation in the walls and ceilings of a home, for example, also operates year after year after year, saving energy in those bone-chilling winter months as well as the hot muggy days of summer. Like renewable solar energy, energy conservation measures continue to deliver their services dutifully year after year. Efficiency measures do the same. Like buying a new energy-efficient refrigerator.

Many local utilities understand the importance of energy efficiency in meeting future energy demands. And many of them even encourage their customers to conserve energy. What is more, some utilities offer fairly sizeable financial incentives to customers for energy conservation measures they undertake, like adding insulation to their attics or purchasing energy-efficient appliances. Why?

By promoting energy conservation, energy companies can reduce their need for additional capacity — building another power plant. That is, they can avoid the need to build a costly new power plant to service a small number of new customers. As a result of your efforts, and efforts of other citizens like you, local utilities also avoid having to purchase additional energy on the "spot market," that is, from companies that have an excess to sell. Buying electricity on the spot market can be extremely costly. The savings can be quite significant.

In sum, then, while energy conservation and efficiency measures are not literally "new sources of energy," the end effect is nearly the same. What is more, they are clean and renewable resources, providing benefits year after year. Efficiency and conservation are also a lot cheaper than developing new sources.

BENEFITS OF ENERGY CONSERVATION

Before we begin our exploration of home energy conservation, let's take a few minutes to recap the benefits of efficiency and conservation and visit a few more not mentioned earlier.

Energy conservation and efficiency measures *reduce fuel use and save homeowners money* — often lots of money. For every dollar you spend on energy conservation and efficiency measures, you can easily reap five dollars per year in savings, according to the US Department of Energy. You can also reduce the size and cost of a renewable energy system, saving once again. Think of energy conservation and efficiency as a reliable alternative to the volatile stock market.

Many energy conservation measures *increase comfort* in our homes by reducing cold spots, hot spots, and chilly drafts. In fact, most comfort issues in homes are caused by poor insulation and inadequate air sealing. Rather than turning up the heat to make a cold, drafty home more bearable, a homeowner can seal up the many leaks in the building envelope with a few tubes of caulk and weather stripping, creating a much cozier interior. And, as just pointed out, efficiency and conservation are cheaper alternatives.

By reducing household utility bills, energy conservation can *free up money for other important activities* — such as building a savings accounts, financing a college education, funding family vacations, and fueling retirement accounts.

Energy efficiency measures in our homes may also *qualify homeowners for larger loans* and, occasionally, for lower mortgage rates. Some lenders, for example, FHA, Countrywide Home loans, and Chase offer such benefits for qualifying energy-efficient homes. Check them out when buying or building a new, energy-efficient home.

Energy efficiency can also *reduce maintenance*. A well-insulated home, for instance, reduces the need for backup heating and air conditioning. This puts less strain on furnaces, boilers, air conditioners, and evaporative coolers, meaning they last longer and require less maintenance — saving you money once again and reducing hassles of having a heating system break down in the dead of winter.

Energy conservation measures are often *easy to implement*. It requires very little skill or training to make a home more energy efficient.

Energy conservation measures can be *brought on line quickly*. A trip to the hardware store and a few afternoons working around the house can make significant dents into your annual energy bill.

Energy conservation measures are often *free or quite inexpensive*. Turning off lights when rooms are unoccupied incurs no expense whatsoever, just a little extra effort on your part. Installing compact fluorescent lightbulbs costs a little upfront, but it saves enormous amounts of money over the long haul.

Energy conservation could *increase the value of your home*. In a study that appeared in 1998 in *The Appraisal Journal*, a publication read by house appraisers, the authors report that the selling price of homes increased by nearly $21 for every $1 decrease in fuel bills the homeowners had achieved through energy efficiency. With the average American home using about $2,100 worth of energy per year (up from $1,300 in 2004), a reduction in the utility bill by one half could increase the home value by nearly $20,000. As energy prices escalate, this figure could go higher!

In an energy-tight world, energy efficiency measures could make your home much *easier to sell*. I expect a huge market for energy-efficient homes in the coming decades.

Energy conservation and efficiency reduce fuel use and thus reduce environmental damage resulting from the production and consumption of conventional fuels (see the sidebar "The Impact of Electrical Use").

HOME ENERGY USE

In North America, homes are huge energy consumers. In the United States, homes account for about one fifth of the nation's total annual energy demand. The energy consumed in homes heats, cools, and lights our domiciles. It also powers a mind-boggling list of appliances and electronic gadgets. While homes vary considerably in their annual use, the average cost comes to about $2,100 per year, a figure that's bound to increase over the coming years. Before you can determine where to apply efficiency and conservation measures, it's important to know exactly where all of this energy is going.

As shown in Figure 2-4, nationwide, heating the interior of our homes consumes the largest portion of the energy pie — about 34 percent. Cooling homes consumes about 11 percent of the energy we use each year. Lighting and appliances (other than the fridge) consume one third of our energy. Water heating consumes 14 percent and the refrigerator consumes about 8 percent of the energy pie.

Bear in mind that these figures represent a national average. Your home may deviate considerably from the average. If you are in a warmer climate — say, southern Arizona — energy consumption for air conditioning will be much higher. The amount of energy required for heating will be quite small. If you live in the far northern tip of Idaho, cooling bills in the summer are something you only wish you had to worry about. Or you may own an extra refrigerator and a freezer in your garage. Unfortunately, during the hot days of summer these appliances must work

overtime to keep their contents cold. This drives up your energy bills. Or you may have a terribly inefficient heating system or a leaky house that requires lots more heat than the mythical "average home."

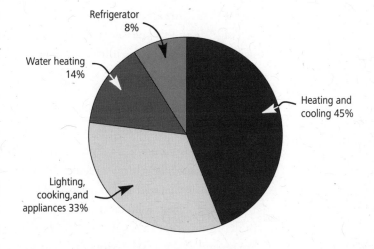

Fig. 2-4:
Average US home energy consumption by end use.

The Impact of Electrical Use

On average, American homes use about 800 to 1,000 kilowatt-hours of electricity each month. All-electric homes consume two to four times as much electricity. On average, each kilowatt-hour of electricity from a coal-fired power plant results in the production of two pounds of carbon dioxide. Electrical consumption in the average American home, therefore, produces about 1,600 to 2,000 pounds of carbon dioxide a year. Cutting grid power electrical consumption through conservation by 50 percent reduces carbon dioxide emissions by 800 to 1,000 pounds per year. Cutting electrical demand in an all-electric home by 50 percent could have even greater benefits.

Although each home is different, this pie chart in Figure 2-4 highlights the big energy consumers. In so doing, it can help us target the greatest potential gains. Valuable as this information is, it is still wise to contact your local utility to see if they can provide an energy pie chart of the "average" home in your area. If your local utility can't provide this information, your state's energy conservation office may be able to offer some assistance.

Obviously, the closer the energy use data you rely on is to homes in your area, the more valuable it will be. Even so, houses range dramatically within a region. Newer homes are often much more energy efficient than older homes. In fact, many homes built prior to the 1970s in the United States are terribly leaky and horribly under-insulated. My students at Colorado College and I, for example, have performed energy analyses on older homes near the campus in Colorado Springs. As part of this analysis, we measure how leaky the houses are (a procedure I'll explain shortly). The test we use simulates a 20-mile-per-hour wind. The houses we test often leak at a rate of 8,500 cubic feet per minute (cfm). (To put this in perspective, my home in Evergreen, Colorado came in at 500 cfm.) The typical rate of 8,500 cfm constitutes a huge leakage problem. Cold air drafts in the winter chill off the interior of the home on windy days. It also results in excessive heat loss at all other times, as warm interior air leaks out through the Swiss cheese exterior of these homes.

Leaky doors and windows can increase the cost of heating or cooling a home substantially, by up to 30 percent, according to Pat Keegan, former executive director of Colorado Energy Science Center, an organization dedicated to promoting residential energy efficiency. "If you think of a house as a boat, a leaky one would sink in an hour," says Keegan.

RETROFITTING YOUR HOME FOR ENERGY EFFICIENCY

Retrofitting a home for energy efficiency will save you money, lots of money. However, it will require some effort on your part. To begin this process, you should start with an energy audit.

Start with an Energy Audit

When retrofitting homes for energy efficiency, as part of the community service component of my classes at Colorado College, we begin with a walk-through, a simple visual inspection of a home. Professional energy auditors do the same.

Starting at the top of the house, we examine the attic insulation, and then take a look at the windows: are they single- or double-paned windows; airtight or leaky? We examine wall insulation, and then search for obvious holes or gaps in the building envelope, for example, broken windows. We then move to the next floor, and then the basement, where many energy problems can be found. You'd be amazed how many we find and how many you will find in your home. Don't forget to check fireplaces; open dampers result in incredible leakage.

After checking out insulation, windows, and leaks in the building envelope, we examine lighting, noting how many lights are in the house and the type of bulb in each one. In the kitchen, we inspect the appliances, especially the refrigerator. Our goal is to determine if there are any old, energy-inefficient models that need replacement. We also examine the water heater. If it appears to be in good operating condition, we note whether it is equipped with an insulation blanket and whether the hot water pipes are insulated. We then examine furnaces and boilers and check out the ducts or pipes that deliver heat (or cool air from a central air conditioner) to the house. We are looking for leaks and insulation — quick fixes that can save a fortune.

These simple visual inspections often turn up huge opportunities for saving massive amounts of energy. In one elderly lady's home, for instance, the class discovered that a heating duct in her uninsulated crawl space had never been connected to the heat register in the living room. The builder had forgotten to install a register, so the furnace installer simply left the job unfinished. When the furnace was running, however, hot air from the orphan duct poured into the uninsulated crawl space, heating the great outdoors. Aghast, my students sealed it off and recommended that the owner hire a handyman to install a heat register in the living room, so the duct could be connected and heat could be delivered to the room.

We make numerous recommendations solely on the basis of our visual inspection. You can perform an inspection of your own home to ferret out energy waste. Just take a pen and paper along with you to record your findings, room by room. Take your time and do a thorough job, especially in the basement, where energy leaks are common. For more advice on performing a home energy audit, you may want to check out my book, *Green Home Improvement*. This book not only offers details on energy audits, it also discusses projects you can undertake to cut energy demand.

Visual inspections reveal the most obvious problems and help us determine the materials and supplies needed to retrofit the home. Once this phase of the home energy audit is complete, we get a bit more scientific. We install a device known as a *blower door* on the front or back door of the house to determine how leaky the house is.

PERRY KIRK

The blower door unit, shown in Figure 2-5, mounts snugly in most doorways. The unit consists of an adjustable aluminum frame covered with Nylon fabric; the unit fits tightly in the doorway. Once the frame and fabric are in place, a large electric fan and an air-flow meter are installed. After the fan is in place, the house is sealed up as it would be in the winter. Exterior doors and windows are shut; attic accesses, if any, are closed, etc. Interior doors are opened so air can flow easily through the house.

When the house is buttoned up, the fan on the blower door test device is switched on. It immediately begins siphoning air from the interior of the house, pushing it outside. The air expelled in this manner

Fig. 2-5: *A blower door test device like this one allows home energy auditors to determine how leaky a house is and to pinpoint the actual leaks so they can be sealed.*

Fig. 2-6: *Air moves in and out of houses through a myriad of cracks, gaps, and holes in the building envelope, as shown here.*

Through vent pipe   Through chimney

DAN CHIRAS

At juncture of roof and walls

Between wall and foundation   Around windows   Around doors

is replaced by air drawn in from the great outdoors through small cracks and large openings, if any, in the building envelope — the foundation, walls, and roof (Figure 2-6).

Once the fan is cranked up, the operator measures the total air flow through the building envelope on the air-flow meter. This measurement is used to calculate the natural air changes — how much air leaks in and out of a home. A good measurement is around 0.35 natural air changes per hour.

Each time we run this test on an old home, we're shocked at how much like Swiss cheese these old homes are. However, I quickly remind my students that there's a silver lining to this black cloud. It is, quite simply, that waste = opportunity. Translated: while our homes are amazingly leaky, embarrassingly leaky, with a little time and money we can seal the leaks. With a little effort, we can dramatically reduce air infiltration and exfiltration (movement of air in and out of a building). It doesn't take much to slice the amount of air flowing through a building envelope by half, sometimes more. The good news, then, is that with $50–$100 dollars worth of caulk and weather stripping, and a couple of afternoons of work, anyone with a lick of sense can dramatically reduce air leakage.

But how does one pinpoint the leaks for air sealing?

We locate many large cracks and openings during the initial visual inspection described above. Obvious cracks and "cat-sized openings" in the foundation — holes big enough for a cat to go through — show up because they let in a lot of light (and cats). To detect leaks, the fan of the blower door device is typically reversed, so it blows air into a home. When the fan in the test device is running, you can detect tinier leaks by feel — that is, by slowly running your hands around window frames, at the junctures of walls and floors or walls and ceilings, at the openings of attic access doors, at outlets and wall switches, and so on. Or you can pinpoint leaks by using incense or a smoke stick. (A smoke stick is a device that produces fake smoke; you can find them in hardware stores.) When the smoke stick is held near a leak, the smoke is blown quickly out through the leak.

By carefully going over the entire house from top to bottom, you can locate the myriad leaks in the building envelope that have been conspiring to make your life miserable — and much more costly than necessary. Although you very likely won't have access to a blower door device, you can feel for leaks on windy days. Take notes so you can come back later to seal up the leaks — or seal them as you go. Be sure to check interior light switches and electrical outlets. A considerable amount of air can enter a home or business through these portals. Outlets and light switches can be sealed by installing specially made gaskets, small rectangular pieces of foam that are installed under the cover plate. You can purchase them in hardware stores or home improvement centers.

Hiring a Professional Energy Auditor

Most homeowners won't want to invest the $3,000 required to purchase a blower door test device, and they're not available for rent unless you are a trained auditor.

The best place to start, then, is with a simple visual inspection to locate the most obvious leaks — gaping openings between doors and door frames, or large openings in the building envelope that let cold air in when the wind's blowing and hot air out when the furnace is running. These can be sealed immediately and will often yield enormous benefits. Then, on a windy day, perform a search and destroy mission for smaller, less obvious leaks, using incense, a smoke stick, or your hand to detect leakage in the building envelope.

If this is more work than you'd like to undertake or if you are feeling uncertain, you can always hire a qualified professional energy auditor to perform the inspection/analysis for you. You'll find them in the business pages under "Energy Conservation and Management Services" or "Home Inspection."

Energy auditors perform complete home energy audits for around $300–$1,000 (depending on the size of your home,) which includes a blower door test, duct leakage tests, and a comprehensive computer energy analysis. Some inspectors even take infrared images of homes to help them pinpoint areas where heat is escaping. They then take this data, along with measurements of the house, and plug it into a computer program to determine how much energy a home is losing.

A good home energy auditor will provide a written report that lists potential energy improvements. The report should

Energy Leakage: Trouble Spots

- Windows and doors: check for leaks around the frames of doors and windows; also check for leaks between the door and the door jamb. Note that many older double-hung and single-hung windows leak.
- Penetrations of the building envelope where water and gas pipes and electrical wires, cables, and phone lines enter, or where dryer vents or vents from kitchen and bathroom fans exhaust to the outside.
- Recessed ceiling lights.
- Attic doors or hatches.
- Whole house fans.
- Electrical switches and outlets on interior and exterior walls.
- The junction of walls and floors along baseboard trim.
- The junction of walls and ceilings.

— US EPA, Energy Star Program

include recommendations for sealing cracks and other openings in the building envelope. If insulation is recommended, be sure that all of the leaks in the building envelope are sealed *before* new insulation is installed. If not, moisture can enter the insulation through cracks in the building envelope. For most types of insulation, even a tiny amount of moisture will reduce the R-value by half.

A professional energy audit will very likely call for additional insulation in walls, ceilings, and perhaps floors, especially over crawl spaces or unheated basements (how much insulation you should add will be discussed shortly).

The report may recommend an insulated blanket for your water heater, and pipe insulation for hot water lines from the water heater to various faucets in the house. In addition, the report will very likely call for efforts to seal and insulate ducts that transport hot and cold air from heaters and air conditioners to the rooms of a house. If your home has a boiler that provides heat for a radiant floor or baseboard water heating system, the auditor's recommendations may call for pipe insulation between the boiler and the various baseboard heaters or heating zones. Lastly, home energy auditors may make recommendations for replacement of old, inefficient appliances. I recommend that once the report is done, you ask for a quick tour of your home so the inspector can show you what needs to be done.

Although a $300–$1,000 energy audit may seem expensive, it is money well spent. This analysis and your follow-up actions can easily pay for themselves in a few years. They'll typically improve comfort right away, too. Check with your local utility company. They may perform low-cost audits, or they may subsidize the cost of the energy audit and energy retrofitting, discussed next.

Who Does the Work?

After reviewing the options outlined in your energy audit, it's time to do something. You have two options: you can do the work yourself, or you can hire an energy retrofitter to perform the necessary work for you. Most auditors, by the way, are qualified to do the energy retrofitting or can recommend someone they trust.

Although hiring a professional will cost substantially more than doing the work yourself, a qualified expert is often worth the extra expense. They perform the work quickly and efficiently, with a minimum of the mistakes commonly made by homeowners. In addition, a trained energy retrofitter will know the products that need to be used, saving you time in the hardware store anguishing over which caulk is best for cracks and which sealant is best for leaky duct work. A professional energy retrofitter can even access high-quality materials that aren't usually sold in hardware stores or building supply outlets, such as duct mastic, a paste used to seal leaky air ducts in heating and air conditioning systems. High-quality products mean a long-lasting job. When hiring a professional, be sure to check references. Also, it's a good idea to hire an experienced professional — someone who has been in the business for at least five years. Watch for the fly-by-nighters. They are out there, especially in areas where local utilities or the local government offer financial rebates! Get references and ask these people specifically about the steps the retrofitter took. Don't hire an energy retrofitter who doesn't insist on sealing leaks before installing insulation.

Other Options for Energy Audits and Upgrades

Rather than hire a professional to perform an audit and retrofit your home, you may want to contact your local utility to see if they can help you out. Many efficiency-conscious utilities not only offer free or low-cost energy audits, they also perform some weatherization work. Years ago, for example, my utility company performed a free energy audit on a rental house I was living in while building my home. In addition to the free blower door test, they gave me two high-efficiency showerheads, three

faucet aerators to save water, and three compact fluorescent lightbulbs — all for free! The auditor even caulked the major cracks in the building envelope at no charge.

Yet another option is to contact non-profit organizations in your area that offer free or low-cost energy audits and retrofits. In Denver, the nonprofit group, Sun Power Inc., offers a wide range of services, including energy audits, insulation, and sealing leaks. They even repair heating and air conditioning systems and install new, energy-efficient heating and air conditioning equipment. Check around. There may be a similar nonprofit group in your area; many of them focus on the residences of low-income families who can't afford high utility bills yet are unable to afford professional energy retrofitting.

Sealing Cracks in the Building Envelope

Sealing cracks in the building envelope is one of the easiest and most cost-effective measures a homeowner can take. And, as I've said a couple times, it's extremely important to seal up the leaks before insulation is installed to protect it from moisture.

For those who want to do the work themselves, I strongly recommend that you purchase the best caulk you can find. Don't skimp on caulk just to save a few bucks per tube. High-quality caulks can last for 20 to 30 years and are worth it in the long run. So ask a knowledgeable worker in the store which products are the best and then spend the extra money.

Caulk comes in three basic varieties: silicone, silicone/latex, and latex. Pure silicone caulk is the best, meaning it lasts the longest. For small gaps, for example, around windows or at the base of walls, I use a high-quality silicone caulk. It comes in clear and white. Larger gaps can be filled with expandable foam, which comes in spray cans. I use Great Stuff insulating foam sealant (Gaps and Cracks — red can) for smaller openings that are too large for silicone caulk. For the largest openings, I use Great Stuff insulating foam sealant (Big Gap Filler — black can). Very large openings can also be filled with backer rod, a flexible, solid tubular material that is pushed into openings. The advantage of foam and backer rod over caulk is that they seal *and* insulate.

In your quest to tighten up your home, you will probably need to seal wall switches and electrical outlets. You'd be amazed how much air leaks into wall cavities and out through attics through wall switches and electrical outlets.

Wall switches and electrical outlets are sealed by installing small foam gaskets; you'll find them in hardware stores and larger building outlets (Figure 2-7). To seal a switch or electrical outlet, even on

DAN CHIRAS

Fig. 2-7: *Foam gaskets like these are used to seal electrical switches and electrical outlets in walls, which can leak quite a lot of air to attic spaces.*

inside walls, remove the cover plate, then insert the foam gasket and screw the cover plate back in place. You can make your own foam gaskets from thin packing foam.

Installing Insulation

In your quest for comfort and energy savings, you will very likely need to add insulation, boosting what's already in place. As a rule, the older your home, the more insulation you'll need to add. If your home was built in the late 1880s or early 1900s, you may find that it has no insulation whatsoever in wall cavities, and barely enough insulation in the attic to make a difference.

How Much Insulation Should You Add?

Professional energy auditors will very likely recommend a boost in insulation to meet local building codes. Although such changes will increase the energy performance of your home and sharply decrease utility bills, I strongly recommend that homeowners go *beyond* the insulation standards in most jurisdictions. In new homes and energy retrofits, I recommend R-50 to R-60 for ceiling insulation and R-25 insulation under floors that are over unconditioned space (unheated or cooled spaces like crawl spaces or garages). Wall insulation should be around R-30.

Insulating to such high levels will cost a few hundred dollars more than merely meeting code, to be sure, but it is worth it in the long run. Remember: energy costs are, very likely, going to continue to climb for a long, long time.

Insulating Wall Cavities

As noted above, many older homes have no insulation at all in their exterior walls. Filling these cavities with insulation is, therefore, vital to achieving comfort and savings. Fortunately, there are several ways to do this. Most often, installers drill large holes in the siding from the outside or in the drywall from the inside, accessing each stud cavity (the space between adjacent studs) individually. They then either blow in cellulose insulation (made from recycled newspapers) or apply a liquid foam product that expands to fill the stud cavity. Some installers blow loose-fill fiberglass into wall cavities, too. The holes used to access the interior of the walls are then repaired and repainted and the job is done.

Wall insulation retrofits can be difficult and are often best left to well-trained professionals with lots of experience. There are many reasons for hiring a professional. First, if your home is two or more stories high, wall insulation may require a lot of potentially dangerous ladder work. Second, cellulose, fiberglass, or (to a lesser extent) liquid foam insulation may hang up on wires and pipes in stud cavities or on lath in older plastered walls. This, in turn, may result in large air pockets in the wall cavities that provide no protection against heat loss. These create cold spots in the winter where heat can leak out. In the summer, heat can enter through these areas. A professional will know how to install insulation without creating air pockets.

Hiring a professional will cost more, as you would expect, but a highly qualified crew can do the job in a fraction of the time it takes most homeowners to complete the task. If the crew is experienced and conscientious, they will also probably do a much better job than the average homeowner.

Although cellulose is a popular product for insulating walls, there are some good alternatives that perform well and are good for people and the environment. One relatively new product is Icynene. Icynene is water-soluble, environmentally friendly liquid foam insulation. It contains no formaldehyde. It also contains no CFCs or HCFCs, both ozone-depleting chemicals, as do other foam products. Icynene is an open-cell foam that seals wall cavities, reducing infiltration and exfiltration. You can't buy the stuff at a local building supply outlet and apply it yourself. It must be applied by a licensed applicator — for one reason, it requires very expensive equipment. Be sure the insulator installs a low-expansion formulation of Icynene in walls and has already had success with it. I'm told that it can be tricky to get right. If the foam expands too much, it can pop drywall off the studs.

Closed-cell (more dense) foams are also available, and provide twice the R-value of open-cell foams. Be sure to look into this option as well.

Another option, rarely pursued, is to remove interior wall material — usually drywall or plaster —then install the insulation in the exposed wall cavity. Damp-blown cellulose, liquid foam, or fiberglass insulation batts, or even loose-fill fiberglass can be used in such instances. When the insulation is in place, new drywall is applied. The wall is then taped and painted. This is a good strategy if you need to replace interior walls damaged by a flood or a water leak. After tearing off the drywall, you may want to deepen the wall cavity. If your home was framed in 2 x 4s originally (a very common practice), a deeper cavity will allow you to beef up the insulation to levels that will make a difference in your comfort and energy savings.

Fig. 2-8: *(a) Author's older brother, Stan, gives a thumbs-up for the energy efficiency measures being taken to remake this building into a classroom for workshops at The Evergreen Institute.*

Fig. 2-8: *(b) The wall in the newly remodeled classroom building consists of two 2 x 4 walls with a two-inch space between, creating a total thickness of nine inches. After exterior sheathing is applied, this space was filled with liquid foam insulation by Sustainable Energy Systems.*

To deepen a wall cavity, you can build a second 2 x 4 wall against the original wall — or a few inches away from the wall — to create a deeper cavity (Figure 2-8). Fill the

Fig. 2-9: *Workers inject a liquid foam insulation that expands to fill the wall cavity in our classroom building.*

DAN CHIRAS

original wall cavity with insulation, then insulate the new wall. Drywall or plaster is then applied to finish the wall.

The double-wall technique dramatically increases the thickness of the insulation layer and dramatically improves the comfort of your home. We retrofitted our classroom building at The Evergreen Institute this way (Figure 2-9). As you can imagine, however, this retrofit is a bit time-consuming and costly unless you do the work yourself, which we did. The insulation was kindly donated by Rocky Huffman, owner of Sustainable Energy Products, an insulation company in Wichita, Kansas. Note that in addition to building a second framed wall, you'll need to increase the depth of the window openings. Baseboards will also need to be removed and reapplied. Wall switches and electrical outlets will need to be remounted. It also slightly decreases interior floor space.

When using the double-wall technique to increase energy performance, be sure to offset the second frame. That is, place the new studs so that they aren't aligned with the original studs. This technique reduces bridging loss — the loss of heat that occurs through the framing members of a wall. Heat conducts fairly rapidly through studs in a wall, because wood offers little resistance to heat flow by conduction. (The R-value of a wood stud is about 1 per inch, compared to damp-blown cellulose, which is about 3.6 per inch). The second wall will not be supporting anything — in other words, it is a non-load-bearing wall — so you can space studs 24 inches on center to reduce lumber costs, reduce bridging loss, and increase the amount of insulation you can put into the wall.

When installing batt insulation into a stud cavity, be sure there are no air spaces between the insulation and the studs.

Table 2-2 R-Values of Insulation		
Material	**R-Value Per Inch**	**Uses**
Loose-Fill and Batts		
Fiberglass (low density)	2.2	Walls and ceilings
Fiberglass (medium density)	2.6	Same
Fiberglass (high density)	3.0	Same
Cellulose (dry)	3.2	Same
Wet-Spray Cellulose	3.5	Same
Rock Wool	3.1	Same
Cotton	3.2	Same
Rigid Foam and Liquid Foam		
Expanded Polystyrene (Beadboard)	3.8 to 4.4	Foundations, walls ceilings, and roofs
Extruded Polystyrene (Pinkboard and blueboard)	5	Same
Polyisocyanurate	6.5 to 8	Same
Roxul (Rigid board made from mineral wool)	4.3	Foundations
Icynene	3.6	Walls and ceilings
Air Krete	3.9	Walls and ceilings

Insulation needs to fit snugly in the stud cavities. If not, air currents can form in the stud cavities, circulating warm air between the inside and outside of the wall, dramatically reducing the R-value of the wall (R-value is a measure of a material's resistance to heat flow). Also be sure not to compress the insulation around pipes or electrical wires. This also reduces the R-value of the wall.

Insulating Ceilings

Ceiling insulation must also be fortified in both old and new homes. Even if your home was built in the 1990s, it will very likely be worth adding more ceiling insulation. Even if the insulation value in the ceiling was previously boosted to R-30 or R-38, increasing the R-value to R-50 or R-60 makes good economic sense. Adding insulation is a breeze in homes with attics. You simply climb into the attic through an access hatch, usually located in the ceiling of a hallway or a back room. Once in the attic, you can lay down fiberglass batts or blow fiberglass or cellulose over the existing insulation. Be sure to walk on the rafters so as not to fall through the ceiling when moving about in an attic. Cellulose is the greenest of these products, as it is made from recycled newspaper.

When retrofitting ceiling or attic insulation, I recommend bringing the R-value well above local building codes — to R-50 or R-60 — in most climates. In really cold climates, you may want to boost insulation even more, up to R-70 or R-80.

When it comes to insulation, many builders in hot climates fall into a trap set by conventional "wisdom." Conventional wisdom mistakenly holds that insulation's main value is in keeping houses warm in cold climates. That is to say, people think it is only important in such climates because it helps to keep heat from escaping. (You don't need a heavy coat in a warm climate, after all, now do you?)

Although insulation is not needed as much in hot climates in the winters, it's absolutely vital in the summer. Why?

Because insulation in walls acts the same way as insulation in a thermos bottle. If you pour a hot drink into a thermos bottle, it stays hot. If you pour a cold liquid in the bottle, it stays cold. Why is this?

Protecting Insulation from Moisture

For optimal performance, many common forms of insulation, such as cellulose and fiberglass, need to be kept dry. This can be achieved by applying a vapor barrier — a six mil layer of plastic applied to the studs after the insulation is put in place. Vapor barriers reduce moisture penetration through the walls of a home. This, in turn, helps to keep the insulation dry, which is important because even a tiny amount of moisture can dramatically decrease insulation's R-value. By preventing moisture from entering wall cavities, vapor barriers also prevent water condensation inside the wall. This usually occurs against the exterior sheathing where the wall is coldest. Moisture condensing inside walls may promote the growth of mold and mildew in the stud cavities — on the framing members and in the insulation. Spores from mold and mildew may enter a home, causing obnoxious smells, discomfort, and even flu-like symptoms. As if that's not bad enough, moisture can cause wooden studs to rot, which can lead to a costly repair job.

In cold climates, vapor barriers are applied to the framing members after the insulation is installed and before new drywall is put in place. In warm climates, however, moisture barriers need to be applied to the *outside* of walls. In new construction, they are applied on the framing members just beneath the exterior sheathing. But that's not possible in a retrofit unless you are planning on installing insulation from the outside — something rarely done. To apply insulation from the outside, you tear off the old siding, remove the exterior sheathing, and then fill the wall cavity with insulation. A moisture barrier is then applied to ☞

Insulation in the thermos bottle prevents heat from moving in *or* out. If you pour hot liquid into a thermos, the insulation prevents heat from escaping. If you pour cold liquid into a thermos, the insulation keeps heat from entering. The same pertains to insulation in homes: in the winter, insulation keeps the interior of homes warm because it prevents heat from escaping; in the summer, it prevents heat from entering, keeping the interior cool even on extremely hot days. It may come as a surprise to many readers to learn that, in many hot climates, utility bills for cooling often exceed winter time heating bills in colder climates.

If you don't have an attic, and your home was built with vaulted ceilings, installing additional insulation can be very difficult, if not impossible. (Figure 2-10).

But all is not lost. You could, for instance, install additional insulation by framing in a second ceiling with 2 x 4s and adding insulation to the cavities. Or you could apply large sheets of foam insulation. Once the insulation is in place, you'll need to attach a vapor barrier and new drywall. Although feasible, this technique is costly the framing members. Exterior sheathing is then reapplied, followed by siding. Needless to say, that's quite a lot of work!

Why do builders install vapor barriers on the inside of walls in cold climates like Maine and North Dakota, but on the outside in warm climates like Florida and Georgia? The reason's simple. In cold climates, the major source of moisture entering wall cavities is interior room air that is laden with moisture from cooking, washing machines, showers, plants and a variety of other sources. Even the occupants of a home release considerable amounts of moisture through perspiration and respiration. In warm climates, moisture that enters walls tends to come from outside the home, from the moisture-laden atmosphere.

While much attention is given to vapor barriers, however, most air enters walls through improper flashing and penetrations. Therefore, be sure that all openings in the roof for pipes are well flashed to prevent moisture from entering your home or business. Also be sure that flashing is applied properly when the roof is installed in valleys and around chimneys, and between roofs and adjoining walls. Also, be sure to seal all penetrations in the building envelope (where wires and pipes enter, for instance) and all interior leaks, including those that occur in light switches and electrical outlets. Be sure to install airtight recessed lighting. These details are often overlooked, even though the vast majority of moisture entering wall cavities from the interiors of our homes enters through such openings.

a   b

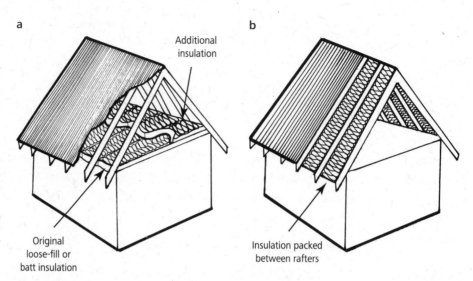

Fig. 2-10: *Two types of basic roof designs are commonly used in homes: (a) the attic design with a large, accessible space that can usually be easily retrofitted with additional insulation, and (b) the closed roof design that consists of an inaccessible insulated space.*

Fig. 2-11: *These 2 x 2 purlins or "sleepers" ripped from 2 x 4s were installed perpendicularly. Rigid foam insulation will be placed between them and metal roofing will then be screwed into them, allowing additional insulation to be added to the roof.*

and time-consuming, especially for home-owners with little experience in this area.

Another option would be to apply rigid foam insulation on the roof. For example, if you need to re-shingle your roof you can begin by tearing off the old shingles. When this is completed, lay down a two-to-three-inch thick layer of rigid foam insulation. New sheathing would then be applied to the foam. Shingles or metal roofing would then be attached to the decking.

Alternatively, you could leave the shingles in place, and attach two-by-four sleepers along the length of the roof at two- to four-foot intervals (Figure 2-11). Foam insulation could be applied between the sleepers. Next, metal roofing could be screwed into the sleepers. This method does not require a new roof deck.

Both of these techniques will greatly boost the R-value of your roof, although, as you can imagine, it is not going to be cheap, nor is it going to be easy. You'll need to pay attention to details like the fascia and overhang. I recommend that you hire a professional to do the work unless you're quite skilled at carpentry.

Upgrading Windows

Windows are a frequent target of energy-efficiency upgrades, especially in homes built prior to the 1980s. Even in many newer homes, cheap energy-inefficient windows represent a huge source of energy loss. Not only are they poorly built and leaky, many windows are improperly installed. As a result, they almost always require improvements or outright replacement.

If your windows are old and leaky or were manufactured with aluminum or steel frames, or, God forbid, are the single-pane variety, you should consider a complete upgrade. The cheapest and easiest upgrade for energy-wasteful windows is to install storm windows. Storm windows can be installed on the outside of existing windows where they perform two vital functions. First, storm windows dramatically reduce air leakage. Second, they create a dead airspace that reduces heat loss. (Remember: air is a poor conductor of heat.) These simple, relatively inexpensive

DAN CHIRAS

Fig. 2-12: *Dan's class in home energy efficiency installed an interior storm window on a residence in Denver to reduce heat loss in a chilly room. The wood trim applied to the sill and around the rest of the opening will create an insulating air space between the new interior storm window and the existing single-pane window glass.*

upgrades can dramatically decrease heat loss and increase interior comfort.

You may also want to consider installing interior storm windows. Now becoming more popular, interior storm windows are relatively inexpensive, attractive, and cost-effective. You can even build your own (Figure 2-12). See the accompanying box for a description of how I make interior storms.

Another, more costly option, is to tear out existing windows and replace them with high-quality, energy-efficient models widely available throughout North America. Most people hire professionals to do this; it's not a job for the untrained, inexperienced, occasional weekend handyman or handywoman.

Window replacements cost quite a lot — potentially $10,000 to $20,000 for a complete replacement, or even more for large homes. The payback period, that is, the time in which this investment pays for itself in energy savings, is typically about 10 to 12 years at current energy costs.

Although that may seem like a long time, a ten-year payback is actually a 10 percent return on your investment. Although this upgrade is costly, the gains in comfort are immediate, and a 10 percent return on your investment is pretty attractive. Moreover, rising fuel bills could help accelerate the payback.

Knowing what kind of windows to buy is a bit challenging. Window technology

Making Your Own Interior Storm Windows

One way to boost energy efficiency of your windows and your entire home is to install interior storm windows. You can do so by installing a second or third pane of glass in the existing window frame. To lower the cost and avoid potential safety issues resulting from the installation of another pane of glass, I've devised another option — one that is much cheaper.

My solution is to install Plexiglas along the inside surface of existing windows — creating an interior storm window. Plexiglas is a clear, durable polycarbonate plastic. It can cost a tenth of the price of some window glass, and won't yellow in sunlight. It also conducts heat more slowly than glass, which results in a warmer surface and thus less heat loss than glass. Plexiglas is also a lot easier and much safer to handle than glass. You are less likely to cut yourself.

To begin, I measure the length and width of the interior window opening, then subtract about one quarter of an inch to allow for expansion and contraction of the Plexiglas. Don't forget this step!

After measuring all of my windows and making the deductions for expansion and contraction, I call in the order to a local plastic supplier. (I'd recommend you try one retrofit, before tackling all of your windows.)

Plexiglas can be purchased at hardware stores. (Ask, as they usually keep it in the back, out of sight.) But for the best price and widest selection, I typically purchase it from a local plastic supplier I found in the Yellow Pages under "Plastics." I prefer three-eighths-inch or one-quarter-inch Plexiglas, not the cheaper one-eighth-inch variety carried by hardware stores, which is much too flimsy and difficult to work with. I've also found that the one-eighth-inch Plexiglas bows when mounted in windows.

To mount the plastic insert, I use wood trim (one-half by one-half-inch). I nail four pieces of stained or painted trim in the window opening (using finishing nails) against the existing window opening. The trim creates an even base against which the Plexiglas will rest. ☞

has advanced considerably in the past 20 years and because of this can be quite confusing. To help avoid confusion, here are some basic guidelines that will help you select an energy-efficient window to replace your old, leaky ones.

In most cases, existing windows should be replaced with double- or triple-pane assemblies. (Triple-pane glass is pretty expensive and should be used on the north side of homes in cold climates where heat loss is most pronounced during the winter months.) To achieve the highest energy efficiency, select window assemblies whose glass is coated with a special, transparent low-e coating ("low-e" stands for low-emissivity). This thin coat of tin or silver oxide retards heat flow through windows. Low-e windows help hold heat inside your home in the winter but also retard heat

---

Next, peel the paper backing off the Plexiglas and place the plastic against the trim base. I then install four additional pieces of trim to hold the Plexiglas in place. In some windows, I've used clear plastic mirror mounts to secure the Plexiglas, screwing the mounts directly to the sash, the wood that holds the glass in place in a window. In some windows, I've found it necessary to screw the mirror mounts to the window frame, that is, the wood that attaches the sash to the rough window opening. (If these terms don't make sense, go take a look at one of your windows. You'll see that the glass is held in place by wood, known as the *sash,* and that the sash is mounted inside the wood frame. It fits into a rough opening created by framing lumber, which you won't be able to see.)

Although windows retrofitted in this manner do not look as nice as those I've retrofitted with wood trim and Plexiglas, this technique is much less expensive and much easier. You won't have to buy, cut, stain or paint, and mount the wood trim. Moreover, this technique makes it easy to remove the Plexiglas from windows you like to open in warmer weather.

Magnetic tape can also be used to mount Plexiglas in windows that you sometimes open. I begin by securing a trim base, as described above. Next, I strip the plastic backing off a piece of magnetic tape cut to fit the length of the trim. I then stick the magnetic tape to the trim base and nail it to the trim using finishing nails. (This gives it a more secure anchorage than the glue on the back of the magnetic strip.)

Once the base pieces are fitted with magnetic strips, I attach a second piece of magnetic tape along the perimeter of the Plexiglas insert. When this is in place, I insert the Plexiglas into the window opening against the base trim. That's it. The Plexiglas can be removed in the summer so the window can be opened for natural ventilation. Unfortunately, the glue on the magnetic tape seems to lose strength over time, causing it to peel off.

Existing skylights can also be retrofitted with Plexiglas to boost their energy efficiency.

flow *into* your home during the summer. As a result, they help maintain year-round comfort.

Also, be sure to select a window with *warm edges*. Warm edges are created by insulated spacers that are placed between the panes of glass along the edge of the window assembly. Spacers reduce conductive heat loss around the periphery of the window, which can be quite significant. This feature not only reduces heat loss, it reduces water condensation on windows on cold winter nights and that, in turn, leads to longer window life.

Another feature to look for is the *U-factor*. U-factor is a measure of heat transmission through a material and is the reciprocal of R-factor (U-factor = 1/R-factor). A window with a U-factor of 0.3 has an R-value of 3.3. Be sure to buy windows with a U-factor of 0.1 to 0.3.

The U-factor of a window is a function of the number of panes of glass, the presence or absence of window spacers, and low-e coatings. It is also determined by the presence of inert gases between the panes of glass. As you shop for windows, you will find that many manufacturers pump the inert gas argon into the air spaces between glass in their double- and triple-pane window assemblies. Argon gas further reduces heat flow, reducing the U-value.

When shopping for windows, look for models that are as airtight as possible.

Resist the temptation to save a few bucks by sacrificing air tightness. You'll be sorry you did. Air leakage is measured in cubic feet of air per minute (cfm) per square foot of window surface. Look for windows with certified air leakage rates of 0.3 cfm/square foot or less.

Another thing to look for when shopping for windows is solar heat gain coefficient — the amount of solar energy a window lets in. Solar heat gain coefficient (SHGC) ranges from 0 to 1. The higher the number, the more light can enter and the more solar heat you'll gain. A high solar heat gain coefficient is extremely important in south-facing windows — to allow the low-angled winter sun to enter so it can heat your home. In temperate climates, a solar heat gain coefficient of 0.5 or higher is ideal for south-facing glass. Since east- and west-facing glass allows a lot of heat in during the summer, it's a good idea to install windows on the east and west side of a home with a low solar heat gain coefficient, which means less than 0.35. So how do you find out about all of these features? How do you know which window is best?

You'll be happy to find that manufacturers post all of this information on their products. Every window comes with a sticker that lists its vital stats such as the U-factor and air infiltration rates (Figure 2-13). The information on these stickers will help you select the most energy-efficient windows possible. You can also get a considerable amount of help from local, knowledgeable window suppliers. (For more details on windows, check out the articles and books on windows in the Resource Guide or read the section on windows in my book, *The Solar House*.)

Instead of replacing windows or installing storm windows, many people tape plastic film over their windows each year as winter approaches. This effective but temporary answer to leaky, energy-inefficient windows, leaves much to be desired. Plastic tends to obscure views and looks … well, let's be honest … cheap. Plus, it requires a lot of work.

Another simple solution to energy-wasteful windows is to put new windows in the existing opening. Contact a professional window installer to see if this might work. It's cheaper and works well if your window frame is in good shape.

Replacing Energy-Inefficient Heating Systems

Sealing leaks in the building envelope, boosting insulation in walls and ceilings, and replacing or retrofitting windows can substantially increase the comfort of your home. They will also dramatically reduce your dependence on costly fossil fuels and even costlier nuclear energy (for homes heated with electricity from nuclear

Fig. 2-13: *This NFRC (National Fenestration Ratings Council) window sticker lists the features of these energy-efficient windows my students and I installed in our new classroom building. These are on east-facing windows, so I selected a low-e window with a low solar heat gain coefficient.*

plants). You can also achieve huge reductions in energy consumption by replacing energy-consuming appliances and other devices — especially heating and cooling equipment, water heaters, refrigerators, and washing machines — with newer, energy-efficient models. (Be sure to recycle your old appliance, if possible.)

When buying replacements for inefficient appliances, seek out the most energy-efficient alternatives on the market. Take furnaces and boilers as an example. In recent years, manufacturers have introduced numerous state-of-the-art energy-efficient furnaces (for forced hot air heating systems) and boilers (for

Fig. 2-14: *Heat pumps like the one on the left in (a) strip heat from the ground via pipes (b) buried below the frost line.*

radiant floor and baseboard hot water systems). Furnaces and boilers use fossil fuels (usually either natural gas or oil) or electricity. The newest furnaces achieve efficiencies in the 83 to 97 percent range, while energy-efficient boilers achieve efficiencies over 90 percent and higher. Efficiencies for new models are posted on a sticker as AFUE (annual fuel utilization efficiency).

Heat Pumps

If your boiler or furnace is old and rickety and begging to be replaced, you'd be wise to retire the dinosaur and purchase a new, energy-efficient model to take over. Another option is to replace your furnace or boiler with a heat pump. Heat pumps are devices used to heat *and* cool buildings (Figure 2-14). As explained shortly, heat pumps remove heat from the environment — from either the air or the ground or a nearby pond, depending on the model — concentrate it and then transfer it to our homes during the winter. Heat pumps deliver heat to existing heat distribution systems, for example, forced-air, or radiant-floor heat, or baseboard hot water.

Heat pumps come in two basic varieties: air-source and ground-source. Air-source heat pumps are usually used in warmer climates, for example, the southern United States, where freezing temperatures are a rarity. But some manufacturers now make air-source heat pumps that work in cold temperatures, down to 10°F.

In an air-source heat pump, heat is drawn from ambient air, even on cold winter days, then concentrated, and used to heat homes and businesses. Yes, you've read that correctly: cold outside air is used to heat a home on chilly winter days. These seemingly magical devices achieve this amazing feat by using refrigeration technology. Basically, as long as the air outside is warmer than the refrigerant in the air-source heat pump, heat will be transferred to the refrigerant. It can then be extracted and used to heat buildings. In the summer, air-source heat pumps operate in reverse, that is, they draw heat out of a house, cooling it down.

Ground-source heat pumps are commonly referred to as *geothermal* systems. That is because they capture and concentrate heat from the Earth. In this system, shown in Figure 2-15, a fluid is pumped through pipes laid in the ground well below the frost line. The fluid circulating through the tubing picks up heat, which the heat pump concentrates and then delivers to the house.

Ground-source heat pumps are remarkably energy efficient. They produce about four times more heat energy than the electrical energy they consume. (Air-source heat pumps produce three times more heat energy than electrical energy they

run on.) Because of their efficiency and the continued wide availability of electrical energy, heat pumps may be one of the shining stars of our energy future. Not only do they heat and cool homes, they also can be used to heat water for domestic use.

Unfortunately, ground-source heat pumps are fairly expensive to add to an existing home. Boring vertical holes or digging trenches to bury the pipe can be costly. Thus, although you'll save a lot in heating costs, you'll pay more upfront. Before you pursue this route, be sure to attend to the changes required to make your home energy-efficient. Pay special attention to the modifications that will

Fig. 2-15:
Ground-source heat pumps, while energy efficient, can be costly because they require extensive excavation. This technology may help us make a transition to a more efficient, renewable energy system, but it won't be cheap.

reduce your heating and cooling loads (that is, how much energy you need to heat and cool your home). You may also want to consider adding passive solar, described in Chapter 4, before contemplating a heat pump. If you live in an area with abundant wood supplies, you may want to consider installing a super-efficient wood stove instead of a heat pump.

Energy efficiency measures, passive solar, and a backup wood stove might be all you need to heat your home. Solar heating options are covered in Chapters 4 and 5.

As a final note, if your furnace or boiler doesn't need replacement or you can't justify the costs, you should call your local utility — or a local heating contractor — to check the efficiency of your system. They may be able to adjust it to burn more efficiently. Also, be sure to replace your forced-air heating system's filter frequently to maintain efficient operation.

Replacing Energy-Inefficient Cooling Systems

Cooling loads (the amount of energy required to cool our homes) are likely to increase in the near future in many places as the world grows warmer, thanks to global warming. The cost of electricity will also increase. (In the United States, the cost of electricity has increased 4.4 percent a year for the past 35 years!)

If you need to upgrade your cooling system, there are two things you should consider: first, be sure that you have made your home as efficient as humanly possible. Seal up the leaks and boost insulation levels. You may need to upgrade windows or install storm windows. Then, read Chapter 7. In it, I discuss the many ways you can cool your home passively — at little or no additional cost. After you've completed all these steps, you may not need to change your existing cooling systems. Your home may stay cool on its own. I've found this to be the case at our educational center in east-central Missouri. The house I stay in at the facility is extremely airtight. I boosted the ceiling insulation to R-65 and took many other steps to reduce heat gain in the summer. As a result, I rarely have to run the air conditioner to cool the house down — unlike residents in the area who run their air conditioners all summer long.

What to do with Your Water Heater?

Water heaters consume a large chunk of the home energy pie. If your water heater is old and decrepit, replace it with an energy-efficient model. You should give strong consideration to an on-demand or instantaneous water heater, described in Chapter 3.

If your water heater still has some good years left, you can cut fuel bills by making a few adjustments. The first thing you should do is check the water temperature. To do this, turn on the hot water and let it run for a minute or two. Next, hold a thermometer in a cup under a faucet for the most accurate reading. If the temperature is above 120°F (49°C), turn the water heater's temperature down. You don't need water to be hotter than 120°F (49°C) to wash dishes and clothes. This is sufficient to kill bacteria. Most people adjust their shower temperature to 104°F (40°C) as well.

If you have a gas- or propane-powered water heater, simply turn the dial on the unit to a lower setting, then wait a day or two and check the temperature again. If the temperature is still too high, lower it again.

For electric water heaters, you'll need to remove the top and bottom covers on the side of the unit, one for each heating element, and turn the temperature settings down using a screwdriver.

Lowering the temperature on a thermostat results in huge savings, but that's not all you can do. After lowering the water temperature setting, you can put an insulated water heater blanket over the tank. Water heater blankets come in several varieties and are available in all hardware stores. I like the bubble wrap plastic variety with metal coating the best. My second favorite is made from aluminum and felt

insulation material. My least favorite is the plastic water heater blanket with a fiberglass insulation interior.

Water heater blankets wrap around the water heater and are secured with tape that comes with the blanket. Follow the installation instructions precisely. Be very sure you do not obstruct the flow of air to the top and bottom of gas- or propane-fired water heaters. Also take care not to cover the dials used to adjust water temperature.

Water heater blankets cost about $18–$20, and pay for themselves in less than a year, depending on your family's hot water consumption. That means every year your water heater blanket is in operation, you'll save $5 to $20 on your annual energy bill. As energy costs go higher, you'll save even more!

After the water heater blanket has been installed, I strongly suggest that you insulate the hot water pipes in the crawl space or basement of your home. I like the foam insulation sleeves that fit over the copper pipe better than the type of insulation you wrap on the pipe. They're much quicker and easier to install.

You can also save hot water — and reduce energy demand — by installing water-efficient showerheads. Most showerheads in older homes consume 3.5 to 5 gallons of hot water a minute. Newer models use around 1.5 to 1.9 gallons per minute and cost about $5–$20, depending on the model. Water-efficient shower-heads reduce hot water use dramatically. One popular model reportedly saves a family of four up to $250 per year. Even if a $10 shower head saves a family of four $100 per year, it's a terrific investment. If only our retirement funds performed as well! As energy prices climb, this simple, inexpensive device could reap even higher dividends.

When purchasing a water-efficient showerhead, I strongly recommend you select one that comes with an on-off button or lever. This feature allows you to turn off the water when lathering up, further cutting back on water and energy use.

You can also save energy and water by installing efficient appliances — notably washing machines and dishwashers, a topic discussed shortly.

Replacing Inefficient Lights

One of the easiest ways to reduce fuel bills is by installing compact fluorescent

Fig. 2-16: In some compact fluorescent lightbulbs, a fluorescent tube is placed inside a glass globe, as shown here. Left side, broken bulb revealing coiled tube inside.

DAN CHIRAS

lightbulbs in commonly used fixtures. And, as you already know, cutting down on electrical use is of great importance to the environment. Reducing electrical energy demand is doubly important for those who are planning on installing a wind generator or photovoltaics (PVs) to produce their own electricity.

I've made the case for compact fluorescent lightbulbs earlier in this chapter, demonstrating how much energy and money they save. CFLs screw into ordinary sockets and can be used almost everywhere you would use regular light-bulbs — ordinary table lamps, ceiling lights, some chandeliers, recessed lighting, indoor spotlights, and even outdoor lights on patios, in garages, and as spotlights on pathways and driveways. Some manu-facturers even produce dimmable CFLs (most CFLs can't be used with dimmable or three-way switches).

CFLs use one fourth as much energy as standard lightbulbs, as noted earlier in the chapter, and give off much less heat, too. (They will help keep a home cooler in the summer as a result.) Each bulb is equipped with a ballast and a small tube, which is typically coiled or looped inside a glass globe (Figure 2-16).

Unlike typical fluorescent lights, CFLs do not produce a spooky cool white light that would make even Jennifer Aniston or Brad Pitt look hideous. Rather, compact fluorescent lightbulbs are coated with special chemicals (fluors) that produce a yellowish light, much like incandescent lightbulbs. They're good for most applications. You can also purchase CFLs with several other types of light, for instance, a bright bluish light.

You can purchase CFLs at hardware stores, building supply outlets, discount stores, drug stores, grocery stores, and even huge warehouses like Costco. It is important to note that it doesn't make sense to install compact fluorescent lightbulbs in every light fixture in your house. It's best to use them in lights that are on for at least three or four hours a day, such as kitchen or living room lamps. It such locations, they will earn their keep, saving you the most money. (Turning a compact fluorescent light on and off frequently will reduce its lifespan.)

If you haven't already installed CFLs in your home, I'd recommend that you buy a couple and give them a try. You may be pleasantly surprised by how well they work. I know I was!

CFLs come in several different shapes and sizes, so they fit into a wide variety of light fixtures and lamps. When selecting a bulb, you will notice that the manufacturers list the actual wattage of each bulb — how much energy each type consumes — and the wattage of the incandescent lightbulb each model replaces. For example, the package of a 23-watt CFL may indicate that it replaces a 100-watt incandescent. In my experience, I've found that it is best to go up a step. That is, if you want to replace a 75-watt incandescent lightbulb with a CFL, go to a CFL designed to replace a 100-watt incandescent bulb.

Another option mentioned earlier are LED lights. They're now starting to enter the market. I use an LED light in the two refrigerators at The Evergreen Institute's Center for Renewable Energy and Green Building. They're 4-watt bulbs that replaced 40-watt bulbs. We also installed a 10-watt LED flood lamp that replaced a 150-watt incandescent flood lamp and have several other LEDs in key areas. LEDs are expensive and some give off a rather cold, blue light. So, shop carefully,

CFL Reading Lamps

At least one manufacturer (Sunwave) produces a line of CFLs designed for reading. Their CFLs emit light that is similar to sunlight at noon, a cooler, bluer light than standard CFLs. I'm working under one right now and have used it for several years. The color of the light isn't as appealing as a standard CFL, but the bulbs do really make it much easier to read. According to the manufacturer, and several years of field testing in my home, these bulbs increase resolution — the ability of the eye to detect detail. According to the manufacturer, they also reduce eyestrain.

and know that the price will come down in time and the quality will improve.

Replacing Inefficient Appliances.

Dishwashers and washing machines may also need replacement. Once again, if your appliances are old and rickety and in dire need of replacement, you'd be wise to go with the most energy- and water-efficient models you can get.

As for washing machines, I strongly recommend the front-loading (or horizontal axis) machines like Frigidaire's Gallery. I use this model and have been extremely pleased with it. It's quiet, and uses one half

the water and a third of the electricity of my old top-loading washer. It did cost more than a standard washing machine, but will pay back the extra cost in a couple of years in water and energy savings. We installed a LG front-loading washing machine at The Evergreen Institute and like it as well. Both the Frigidaire and LG were well priced — meaning significantly less expensive than other brands we looked at. All major appliance manufacturers now offer front-loading washing machines.

You should also consider replacing your old dishwasher with an energy- and water-efficient model if yours needs replacement. Or — and don't e-mail me in protest over this one — you could do the dishes by hand!

Dishwashing by hand saves a lot of energy and water — if you do it right. Although dishwashers can be just as efficient as washing by hand, most people who use dishwashers rinse their dishes so thoroughly *before* they put them in their dishwashers that they're practically clean *before* they push the "on" button. In the process, they waste a lot of water and energy. In such instances, it would be hard to argue that a dishwasher uses less water than hand washing.

In our home, I wash and rinse dishes in two tubs, one in each half of the divided sink. Each tub holds about a gallon and a half of water. When the dishes are done,

Energy-Efficient Hot Tubs?

When many people think about making their homes energy efficient, they run into a stumbling block when it comes to one item: the hot tub. Electric heaters in hot tubs consume huge amounts of electricity. However, one company, Softub, has come up with an innovative new hot tub design that uses only a fraction of the energy of a standard hot tub. Instead of a standard resistance-type heater, the water is heated by waste heat from the water circulating pump. Water from the hot tub flows through pipes around the pump, drawing off waste heat that's used to bring the hot tub up to temperature. All in all, they use about $10 worth of electricity a month, compared to a standard hot tub that can use as much as $60. Moreover, their hot tubs are made of foam and are very lightweight. No need to hire a crane to lift it into your backyard. To learn more, visit softub.com we use the relatively soapless rinse water to water indoor plants. The net effect is that dishwashing consumes about two to three gallons per load.

End of sermon.

Where Do You Go for Information?

Buying energy-efficient appliances and heating and cooling equipment requires patience and perseverance. The job, however, is made a lot easier by consumer labels required by the US and Canadian governments. These yellow Energy Guide labels list pertinent energy data on each model — like how the model you are looking at compares to others in its category, how much energy the device will use, and what it will cost you in a year's time. These labels are designed to help consumers make the wisest choices.

EPA Energy Star labels will also help guide you to the best choices. In the US, Energy Star labels, shown in Figure 2-17, are posted on a wide variety of electronic equipment and appliances. They don't list any pertinent data, but serve to indicate the most energy-efficient products in each category. Shop very carefully though. Even among the Energy Star-qualified models, you'll find significant differences. In other words, some are a lot more efficient than others.

To shop for an appliance, visit energystar.gov, select the appliance category,

Fig. 2-17: *The Energy Star label on appliances or electronics indicates that you are looking at one of the most energy-efficient models in its class.*

then click on the list. You'll be given a very large list of options with tons of data to help you select the size appliance you want and its efficiency. You'll even be able to get the model numbers to help you order the exact model you want. Rather than having to shop around, asking ill-informed sales people about their most efficient models, I found the model I wanted on energystar.gov, then called around to see who carried it or who could order it for me. A few phone calls later and you're done.

Ghost Busters

You probably don't know it, but your house is haunted. Yes, haunted. Not with spirits of people who have lived in your home before you, but by ghost loads, or,

Fig. 2-18:
Ghost, or phantom, loads are everywhere in our homes, as in this stereo system. Finding ways to reduce them will help you save energy and make your home more suitable for a renewable energy system.

more commonly, *phantom loads.* Phantom loads are the electricity used by appliances and many electronic devices when they're not in active use. Your instant-on television, for example, is haunted by a phantom load. That is, it consumes electricity even when it's off. Why? To keep the circuits buzzing, so that when you click the power button on the remote, the TV hums to life instantly.

Other examples include the power transformers for cordless phones, answering machines, and keyboards; microwave ovens and coffee makers equipped with LED clocks; stereos, radios, audio receivers, satellite receivers, and power strips (Figure 2-18); even hard-wired smoke detectors and some GFIs (the electrical outlets with built-in circuit breakers). And don't forget your cell phone or notebook computer recharger; it may draw power even when the phone or computer aren't hooked in to them.

Your home is full of phantom loads that, although tiny on their own, add up and collectively consume significant amounts of electricity over a year's time. When I first moved into my solar home, I discovered that there were about 125 watts of cumulative ghost loads. That's the equivalent of two 60-watt lightbulbs running 24 hours a day, 365 days a year. In my case that came to 1,095 kilowatt-hours per year that were putting a huge strain on my tiny solar electric system. Had I been buying power at the local rate, about 8.5 cents per kWh, it would have cost me $93 a year just to feed these annoying ghosts!

Needless to say, I set out to eliminate the ghosts from my life, and so can you. To do so, first look around you. You'll find them everywhere. Make a note of each phantom load, and then devise a strategy to get rid of them one by one. For example, you can plug televisions, stereos, microwaves, and the like into power strips that can be shut off when the electronic device or appliance is not in use (Figure 2-19).

For appliances such as microwaves, you can replace the standard outlets they're plugged into with switch and receptacle outlets. These are electrical outlets, available at all hardware stores, that are controlled by an integrated switch. Turning the switch off, terminates the flow of power to the electrical outlet and thus ends the flow of energy that feeds the ghost load of the microwave or other devices plugged into the outlet. If you aren't skilled in electrical work, I strongly suggest that you enlist the help of a qualified friend or hire a licensed electrician. Be sure to turn the power to the circuit off before you mess with any electrical switch or outlet. Also be sure to buy a switch and receptacle that doesn't come with an LED light, which consumes power 24 hours a day.

Another source of ghost load are GFIs — those electrical outlets in bathrooms that have built-in circuit breakers to protect you from getting shocked if someone drops a hair dryer in the tub while you're taking a leisurely bath. GFIs, technically referred to as GFCIs, for ground fault circuit interrupters, short out before you do. Unfortunately, some GFIs draw power even though there's nothing plugged into them. To see if a GFI has a ghost load, plug in an amp meter. A trained electrician can do this for you. I replaced my GFIs with a no-ghost-load GFI from Leviton.

To reduce energy-draining ghost loads, you can also unplug power transformers or chargers when not in use. For example, if

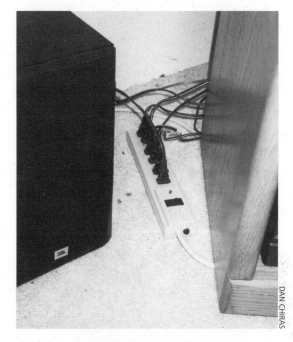

DAN CHIRAS

Fig. 2-19: *A power strip can be used to control phantom loads, but be sure to turn it off when the attached electronics are not in use; even the power strip comes with a phantom load. You can plug your computer, printer, monitor and even your modem and router into a power strip and shut them off at night or when not in use to save energy.*

you have an electronic keyboard, unplug it when you're not using it. Unplug your cell phone charger when it's not in use, too. To learn more about phantom loads, pick up a copy of my article, "Hunting Phantom Loads," which was published in *Home Power* magazine in May 2001. I have also included a project on assessing and getting rid of phantom loads in one of my newest books, *Green Home Improvement*, which contains 65 home projects you can do to green up your act!

THE SILVER LINING

Energy efficiency and energy conservation are vital to each of us and to our collective long-term future. So far, though, North Americans haven't taken energy conservation or efficiency very seriously. Our wastefulness has hastened the depletion of oil and natural gas and seriously dimmed the lights along the path into our once illustrious future. Had we developed an energy efficiency ethic earlier in our history, or even heeded warnings about energy shortages in the 1970s and learned to use fossil fuels efficiently, this book probably wouldn't have been needed for another several decades.

But we didn't and there's no sense in crying over spilled oil. The good news is that waste is the silver lining of the cloud that darkens our future. Because we do waste so much energy, the easiest and most cost-effective way to meet our near-term energy demands is by practicing the lost art of energy conservation and by implementing energy efficiency measures. In so doing, we could reduce potential hardship, making it easier on ourselves, our family, our neighbors, and very likely all of humanity.

You can do your part to help smooth the rocky road to a renewable energy future by spending some time energy retrofitting your home. The sooner the better. Why not head down to the hardware store now and pick up some weather stripping, caulk, and insulation to improve the efficiency of your home? While you are at it, pick up a dozen CFLs and an LED or two, install them, and sit back and enjoy the savings.

As noted in this and the previous chapter, financial assistance from your local utility or your state or local government — even the federal government — can sweeten the deal. A quick phone call to your utility or to your state energy office may reveal that one or the other offers incentives — sometimes fairly sizable financial incentives — to help you become a model of energy efficiency. Federal subsidies and incentives may also be available. You can use your savings to implement other steps in this book, for instance, the installation of solar electric modules; local utilities may provide incentives for this, too. To learn about incentives, visit the Database on Efficiency and Renewables (dsireusa.org), click on your state, then click on the various programs you want to research.

I advise you to get moving now, though. Don't dally. The cost of energy retrofitting will very likely escalate in years to come, as more and more people see the need for it (increasing demand) and as energy prices themselves go up. Remember: it takes energy to create all the energy-efficient products I've been talking about, including energy-efficient appliances, so as the price of energy goes up, so will the cost of everything else, including the devices we buy to reduce our demand for energy.

Nancy Newhall, an American photographer and writer best known for writing the text to accompany photographs by Ansel Adams and Edward Weston, wrote that conservation is "humanity caring for the future," and she's right. Let's not forget that the future starts tomorrow. If we're to have a decent future, we need to get serious about energy conservation today.

CHAPTER 3

Solar Hot Water Systems

SATISFYING DOMESTIC HOT WATER NEEDS WITH SOLAR ENERGY

Solar energy is a gift from the sun, but one that's vastly underutilized by human society. As natural gas supplies dwindle and prices skyrocket, however, more and more people will turn to the sun to meet their needs for energy. We'll use the sun's energy to heat our homes (Chapter 5) and to produce electricity to power homes, cars, schools, hospitals, businesses, and factories (Chapter 8). Many people will also start using the sun's massive outpouring of energy to heat water for showers, baths, washing clothes, and dish washing. This chapter explores this application, known as *domestic solar hot water* (DSHW). Besides ensuring our comfort and promoting personal hygiene, hot water constitutes a huge portion of our annual fuel bill — about 13 percent of the average annual household

fuel bill in the United States, according to the US Department of Energy. As natural gas production declines and prices climb, the cost of supplying domestic hot water could rise meteorically. Those who retrofit their homes for solar hot water could save a sizeable amount of money.

However, before deciding to install a solar hot water system, you should first understand how conventional hot water systems work, because solar hot water systems are typically integrated with them.

CONVENTIONAL HOT WATER SYSTEMS

In most homes in North America, hot water is provided by a conventional gas, electric, or oil storage water heater — widely (and incorrectly) referred to as "hot water heaters." (They don't heat hot water!) The storage water heater consists of a free-standing water tank that holds 40 to 80 gallons of hot water (Figure 3-1). The tank itself consists of a steel tank lined internally by a thin layer of glass (to prevent the steel tank from rusting). The internal steel tank is encased in a steel envelope with a layer of insulation in between them to hold heat in.

In electric storage water heaters, water is heated by two electric resistors, known as *heating elements*. The heating elements — one near the top and one near the bottom of the tank — generate heat when electricity flows through them, much like the electric heating elements of an electric stove. The heat produced by the resistors is transferred to the water in the tank. Each element in an electric water heater has its own thermostat. The lower element, or standby element, located at the bottom of the tank, maintains the minimum thermostat setting; the upper element, the demand element, provides hot water recovery when demand heightens — that is, it heats water very quickly when the demand for hot water increases.

Gas-powered storage water heaters have a single source of heat, a burner at the bottom of the water tank. The burner is connected to a thermostat and ignites when the temperature of the water inside the tank drops below a predetermined set point. When this occurs, the temperature sensor sends a signal to a valve that regulates the flow of natural gas or propane into the burner. When the valve opens, gas flows into the burner and is ignited, most often by a small pilot light that runs 24 hours a day, 7 days a week, 365 days a year. Oil-fired water heaters operate similarly; however, they contain a power burner that mixes oil and air to create a mist that is then ignited by an electric spark.

Because the combustion of natural gas, propane, and oil produce carbon dioxide, carbon monoxide, and nitrogen dioxide,

Fig. 3-1: *Water heaters like this one on right are popular in North America. They store 40 to 80 gallons of hot water, ready for use any time of the day or night. Although they work well, they're not as efficient as other systems. Standby losses account for about 20 percent of the energy consumed by these units.*

ECONAIR

which are potentially poisonous to people and pets, gas-powered water heaters must be vented. Venting is accomplished by a flue pipe that directs hot combustion gases and pollutants through the ceiling. In most gas-fired water heaters, combustion air that feeds the burner comes from room air. For a discussion of newer, safer, and more efficient power-vented water heaters, see the accompanying sidebar, "Power Venting: Induced Draft Water Heaters."

Conventional storage water heaters maintain a large quantity of hot water day in and day out. This water can be used at any time. It's there at our command. Just open a hot water faucet or turn on an appliance like a clothes washer, and you've got hot water in 10 to 30 seconds, depending on the distance between the faucet or appliance and the water heater.

Hot water drained from the tank is replaced by cold water from a cold water

Power Venting: Induced Draft Water Heaters

Standard water heaters are popular in North America, but they are not as efficient or as safe as they could be. To address this, some companies sell power-vented water heaters, typically referred to as *induced draft models*. An induced draft water heater consists of a large storage tank and a burner, just like a conventional gas-powered storage water heater. However, that's where similarities end. This new and safer water heater also includes a fan. It draws outside air into the combustion chamber and forces combustion gases out through a vent. This is important for two reasons: Bringing outside air in helps prevent leakage through cracks in the building envelope that reduces the overall energy efficiency of a home. (Air must enter a home to replace the air that exits with the combustion gases through the flue.) Power venting also actively expels combustion gases and the pollutants they contain. This ensures that pollutants produced by the combustion of natural gas or propane are vented to the outside

and can't enter the room air, causing health problems. Spillage is also prevented by a closed combustion chamber. It contains the fire and the combustion gases, unlike a standard water heater, making it impossible for them to escape into our homes. Power-vented water heaters are not only safer, they're more efficient — usually at least 10 percent more efficient than standard gas-fired storage water heaters. In fact, they're so much more efficient than standard storage water heaters that they can be vented with plastic flue pipes rather than the metal flue pipes of conventional water heaters. (The induced draft water heater removes much more heat from the combustion of natural gas, meaning the combustion gases are cooler.)

While safer and more efficient, these models do cost more, and they use electricity to run the fan. Your gas bills will go down, but your electric bill will go up a tiny bit. It's not a big deal, however. The increase will hardly be noticeable.

line. As it enters the tank, the cold water cools the water in the storage tank. When the temperature of the water in the tank drops below the desired setting, the gas burner ignites. Once the water reaches the desired setting, the flame turns off.

Pros and Cons of Storage Water Heaters

Conventional storage water heaters are widely available in North America, fairly inexpensive, and are about 80 percent efficient. However, they do have some drawbacks. For one, they generally don't last very long. You can count on replacing your water heater every 10 to 15 years, unless you take steps like those outlined in the accompanying sidebar "Water Heater Revival."

Another problem with storage water heaters is that they use natural gas and propane, two fuels slipping toward extinction. Even electric water heaters have some problems. For one, they rely on an expensive source of energy: electricity. (Electricity is the most expensive means of heating water, bar none!) Perhaps the most significant problem with storage water heaters is that they maintain a large quantity of hot water 24 hours a day, 7 days a week, 365 days a year. In between periods of demand, heat leaks out of tanks and the pipes that attach to the top. This is called *standby loss*. Standby heat loss must be replaced. One fifth of the fuel consumed by a conventional storage water heater is used to offset the standby losses.

Many, myself included, believe this system wastes too much energy. It's a little like keeping your car running in the garage 24 hours a day just in case you wake up in the middle of the night and want to take it for a spin! A far better alternative is the tankless water heater.

TANKLESS WATER HEATERS

Travel through Europe and other parts of the world, like Japan, where energy is

Water Heater Revival

Storage water heaters can be made to last much longer and operate more efficiently by draining sediment from the bottom of the tanks every year. Simply open the faucet at the bottom of the tank and drain off a couple of gallons of water containing sediment. You can also extend their working life by replacing the anode in the tank. The anode is a long, slender metal rod that's usually attached to the cold water inlet pipe of the water tank at the top of the tank. It may be attached separately, and is usually well labeled.

The anode rod extends down into the tank and prevents corrosion of the steel casing (for reasons beyond the scope of this book). Periodic replacement can double the working life of a water tank, saving you lots of money. For details on how to replace the anode, I refer you to Chuck Marken's article in issue 106 of *Home Power* magazine, "New Life for Your Old Water Heater" or my book, *Green Home Improvement,* Project 33.

expensive and efficiency is a cultural norm, and you won't find storage water heaters like the ones in North American homes. In such regions, the tankless water heater is the technology of choice (Figure 3-2a). Why? Because tankless water heaters outperform their conventional counterparts easily by 20 percent or more. Like storage water heaters, tankless water heaters provide hot water on demand, but they do so without storing huge quantities of hot water. In fact, as their name suggests, they have no storage capacity at all.

Anatomy of a Tankless Water Heater

Tankless water heaters for household use are attractive, compact units that mount on the wall in a central location or near the main hot water demand centers (Figure 3-2a). The heart and soul of all tankless water heaters is a device called a *heat exchanger* — a combustion chamber in which cold water is quickly brought up to temperature (Figure 3-2b). The heat exchanger transfers heat from the burner to the water, heating it instantaneously. A flue pipe vents unburned gases and pollutants like carbon monoxide out of the house.

Here's how a tankless water heater works: when hot water is required — for example, when someone turns on a hot water faucet — a water flow sensor in the water heater sends a signal to a central

a

Fig. 3-2: *(a) Tankless water heaters like this one heat water instantly (there's no need to store hot water for intermittent uses) and thus reduce the amount of energy required to supply a family with hot water by about 20 percent. These units are typically mounted on walls in a central location in the house. (b) Cutaway of a tankless water heater.*

TAKAGI

b

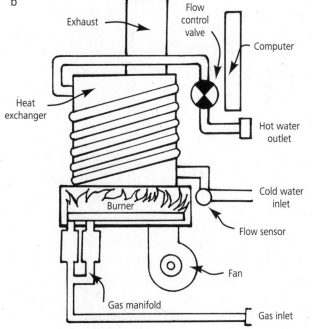

HOME POWER MAGAZINE

Under-the-Sink Water Heaters

Many tankless water heaters in use in Europe fit under the sinks of homes and apartments. They heat water for one sink only and are typically powered by electricity. The household-sized units described in this chapter are centrally located so they can service all hot water needs in a home. They are often powered by natural gas or propane, both of which are much more efficient fuel options than electricity.

control module. This tiny computer, in turn, sends signals to an electronically controlled gas valve in the gas manifold. The valve opens and allows natural gas or propane to flow into the combustion chamber of the heat exchanger. The gas is ignited by a pilot light or by a spark from an electronic ignition device. Cold water flowing through the pipe in the heat exchanger is immediately brought up to the desired temperature. As soon as the hot water faucet is turned off, the water flow through the water heater ceases and the flame goes out. The result is that you heat only the water you need.

Pros and Cons of Tankless Water Heaters

Tankless water heaters cost more than their conventional counterparts and are a bit more challenging to install. (This isn't a job for most do-it-yourselfers.) However, they do provide significant advantages over storage water heaters. As just noted, they heat only the water needed at any one moment. As a result, they are usually at least 20 percent more efficient than standard water heaters. That means they provide the same amount of hot water as a storage water tank, but use 20 percent less fuel. Utility bills will be 20 percent lower, too.

Because there's no standby loss, they also produce less waste heat. In the summer, that means less heat is generated in your home, which lowers cooling bills, the topic of Chapter 7. Another huge advantage of tankless water heaters is that they outlast conventional water heaters. They're typically designed to last as long as a conventional furnace or boiler — twenty years or more. Moreover, tankless water heaters are easy to repair. If a part goes bad, it can be replaced. When something goes wrong, you don't have to throw the unit out and replace it with a brand new water heater, as you do with conventional storage water heaters.

With this background in mind, let's turn our attention to solar hot water systems and how they are integrated with conventional or tankless water heaters. (If you want to learn more about tankless water heaters, I suggest you read Jennifer Weaver's piece on them in *Home Power*, Issue 105 or my book, *Green Home Improvement*, Project 34.)

WHAT IS A SOLAR HOT WATER SYSTEM?

Domestic solar hot water systems require solar collectors, which are typically mounted on the roofs of homes or businesses. The solar collectors capture solar energy to heat water. Most domestic solar hot water systems are designed to provide 40 to 80 percent of a household's annual hot water needs, although 100 percent is possible. (This requires using the most efficient solar panels and is typically achieved in the sunniest locations, like southern California, Arizona, and Florida.)

Domestic solar hot water systems are usually integrated with conventional storage or tankless water heaters. (Be sure to purchase a tankless water heater that's designed to operate in conjunction with a solar hot water system.) As you shall soon see, when operating in conjunction with a solar hot water system, conventional storage or tankless water heaters typically become secondary heat sources. That is, they back up the solar hot water system.

Although this chapter deals with systems for domestic hot water, these systems can also be designed to provide space heat for homes. Solar hot water systems can also be designed to heat greenhouses and swimming pools. They can even provide hot water for hot tubs, saving homeowners tons of money and dramatically reducing the environmental impact of using conventional heat sources. I'll describe solar hot water systems for space heating in Chapter 5. (See Bob Owens's piece on solar hot water systems for hot tubs listed in the Resource Guide.)

But that's not all. Solar hot water systems are also installed for what may seem like a frivolous need — melting snow on driveways. In the upscale ski town of Aspen, Colorado, for example, many wealthy homeowners install solar hot water systems to de-ice their driveways. They're required by law to install some kind of renewable energy system to partly make up for the large homes they build, and these systems sastisfy that requirement.

A BRIEF HISTORY OF SOLAR HOT WATER

Solar hot water is not a new technology by any stretch of the imagination. In fact, solar systems were quite popular in the early 1900s in the United States, particularly in California and Florida. An estimated eight million systems were installed on the rooftops of homes in these areas in the early 1900s.

But along came cheap natural gas, and the solar hot water industry took a nose dive. It wasn't until the oil embargoes of the 1970s that the solar industry got back on its feet again. To decrease our nation's dependence on foreign energy sources, in part by encouraging renewable energy, the federal government and many states offered generous financial incentives — tax credits and tax deductions — for homeowners who installed solar hot water systems. In my home state of Colorado, the combined federal and state incentives covered 60 percent of the cost of a new system.

The domestic solar hot water (DSHW) industry boomed in the late 1970s and early 1980s, but during the Reagan Administration the incentives came to a screeching halt. Neither Congress nor the president was willing to extend them. President Reagan even took down the solar hot water collectors that his predecessor,

Solar Driveway

In Aspen, Colorado, and other areas, solar hot water systems are often installed to keep driveways free of snow and ice. In these systems, pipes are installed beneath the driveway before the pavement is laid down. These pipes circulate a solar-heated fluid (a type of antifreeze) beneath the driveway. Even on cloudy days, the panels often produce enough heat to melt snow and ice, as the fluid circulating beneath the driveway doesn't have to be as hot as water in our homes. It has to be just hot enough to melt snow. This application may seem like a luxury to most of us, and it is, but it beats the alternative: using a gas-fired boiler or electrical wire to achieve the same result. And it does eliminate the need to plow driveways, an activity that uses a lot of energy!

Jimmy Carter, had installed. And so ended the second era of solar hot water.

As Greg Pahl writes in his book, *Natural Home Heating*, "The fall of the industry was as sudden — and spectacular — as its rise." The second coming of solar hot water was not a complete bust, however. Its meteoric rise introduced many to a sensible and economical alternative to electricity, propane, oil, and sometimes even the cheapest fuel, natural gas. Even more importantly, it resulted in the installation of numerous systems, many of which are still operating today.

The rise and fall of the solar hot water industry in the late 1970s and early 1980s did create some lasting problems. Generous tax incentives created an almost overnight industry. In their rush to deliver their products to market and tap into the incentives, some companies rushed products to market before they were fully tested. Moreover, some vendors and installers engaged in unscrupulous activities. Some overstated the economic benefits and many engaged in price gouging. I fell victim to price gouging myself. I bought a system from a company (no longer in business), paying a premium price for the equipment and installation — a whopping $6,000. Like other customers, I was willing to pay what seemed like an outlandish price. Who cared what the sticker prices was? Federal and state incentives knocked the price back down to about $2,400, including installation.

Unfortunately, when the tax incentives withered away, so did most of the companies. Many homeowners, myself included, were abandoned. When systems needed repair, we had no one to turn to. The service personnel were gone, and so were the parts we needed. Many systems fell into disrepair and were removed from roofs. You may even be able to buy some solar hot water collectors removed from roofs during this time. I have two of them myself that I'm going to install at The Evergreen Institute.

Since then, the solar hot water industry has struggled to gain a foothold. But after years of struggle, solar hot water systems are gaining in popularity thanks in part to high fuel prices, and enlightened public policy at the state and federal level. Homeowners and business owners can receive a 30 percent federal tax credit for solar hot water systems. Some states and utilities offer rebates or incentives as well. To learn about rebates in your state, visit dsireusa.org and click on your state on the US map. Even without financial incentives, DSHW systems often make good sense, as you shall soon see.

Despite the resurgence in solar hot water, the solar industry still hasn't completely overcome the bad image lingering in the minds of many people. Unfortunately,

this technology is still viewed by some as unreliable. "Happily, the industry has matured [and] the technology has improved," notes Pahl. Today's companies produce an excellent product. Making things better for homeowners, there are now performance standards for most of the components of active solar heating systems, which makes comparing different products much easier.

I believe that this time solar hot water is here to stay. Incentives offered at the federal and state levels could boost the industry dramatically, but they may not be needed. Market forces may drive the switch to solar hot water, creating a lasting energy source throughout the 21st century and beyond.

My recommendation to those thinking of installing a system is to go with established companies — installers and manufacturers who have been around for five to ten years. "Before you decide to buy an active solar heating system," notes Pahl, "be sure to check local zoning ordinances, land covenants, and any other possible local restrictions" that might apply to you. "Homeowner association rules, in particular, can be a real headache when it comes to solar collectors." Although such restrictions are not widespread, they can be insurmountable. If you do run into local restrictions, you may want to invest some time, energy, and money to reverse them.

As Pahl notes, "Most of these restrictions are absurd and deserve to be changed, so it's probably worth the effort if you are committed to solar heating." We need solar pioneers who will help pave the way for others.

SOLAR HOT WATER SYSTEMS

Domestic solar hot water systems consist of several basic components. The two most prominent are the solar panels (more appropriately referred to as "solar collectors") and the water storage tank. The solar collectors are usually located on the roof or on the ground next to a house or business. In both locations, they need to be positioned so they receive full sunlight year round (Figure 3-3). The water storage tank is typically located next to the conventional water heater, often in the basement of a home. Copper pipes connect the collectors with the storage tank, and various pumps, sensors, valves, and controls ensure that the system works automatically. Because many DSHW systems utilize pumps, they are classified as *active systems*. (As you shall soon see, though, there are a few DSHW systems that require no pumps at all.)

Solar hot water systems work in all climates, from the sunniest areas (the sunbelt) to the dreariest of all climate zones (the so-called Gloom Belts). The type of system you install depends on the climate, as you shall soon see. With this overview in mind, let's take a look at your options, starting with the simplest of all systems.

Table 3-1 provides a summary of the types of systems. You'll note that the table lists whether a system is active or passive, the type of heat transfer fluid that circulates through the solar collectors, what moves this fluid, whether the system is open or closed (explained shortly), and finally, the suitable climate.

SOL-RELIANT

Fig. 3-3: *The most prominent part of a solar hot water system is the solar collector. Collectors are typically mounted on the roofs of houses or alongside homes to ensure year-round exposure to the sun from 9 a.m. to 3 p.m. each day. Heat energy gathered from sunlight is stored in a water tank, usually located in the basement, for use when needed. This model by Sol-Reliant uses a PV module to provide electricity to pump. No additional wiring is necessary.*

Domestic solar hot water systems fit into one of two broad categories: direct or indirect. Open systems (a.k.a. direct systems) heat water that you use. In other words, water circulates through the collectors, is heated and then used inside a home or business. Closed-loop systems (a.k.a. indirect systems) heat a fluid that then heats the water.

Solar Batch Hot Water Systems

The simplest of all the solar hot water systems on the market is the solar batch hot water system. Solar batch water heaters are popular in many tropical countries, such as Mexico. In such locations, solar batch water heaters consist of a single water tank mounted on the roofs of homes and businesses. The tanks are typically painted

Table 3-1 Solar Hot Water Systems					
Type of System	Active or Passive	Heat transfer fluid	Propulsive force for heat transfer liquid	Open- or closed-loop	Suitable climate
Solar batch water heater	Passive	Water	Line pressure	Open	Warm, very infrequent freezing or cold weather
Thermosiphon	Passive	Water or propylene glycol	Convection	Open- or closed- with propylene glycol	Warm, infrequent freezing or shut off in winter
Pump circulation (Gravity drainback)	Active	Water	AC or DC Pump	Open	Any climate but designed for cold climates
Pump circulation (closed-loop antifreeze)	Active	Propylene glycol	AC or DC pump	Closed	Any climate

black to increase the absorption of solar energy.

Solar batch water tanks absorb sunlight all day long, which heats the water inside the tank. When hot water is required, it flows directly out of the top of the tank into the hot water supply line that feeds various hot water faucets inside the building. Cold water enters the bottom of the tank to replenish hot water drawn out of the tank with each use. Because the water used in the house is heated in the tank, the solar batch hot water system is considered a *direct* or *open-loop* solar hot water system.

Solar batch water heaters like these are simple, economical, and reliable. However, they do have some shortcomings. One of those shortcomings is that the tanks are not insulated. As a result, they tend to lose a lot of heat at night. The hottest water is, therefore, typically available only in the afternoons and the early evenings.

But don't close the book on solar batch water heaters just yet. More efficient —

Fig. 3-4: *The solar batch water heater is one of the simplest and most cost-effective solar hot water systems available on the market today. Unfortunately, it can only be used in warmer climates, where freezing rarely, if ever, occurs.*

BOB BOWENS

and more attractive — models are available. One type, shown in Figure 3-4, consists of a large black storage tank inside a glass-covered, insulated collector. Sometimes referred to as *integrated collector and storage* (ICS) *water heaters* because the collector and storage tank are one unit, these models are mounted on roofs, so long as the roof can support the additional weight; they can also be installed on the ground, alongside buildings.

For best results, batch heaters must face true south, not magnetic south. (True south corresponds to the lines of longitude and is not often the same as magnetic south, which is determined by magnetic fields. Magnetic fields don't always run true north and south.) Like all other solar collection devices you'll encounter in this and other chapters, solar batch water heaters need to be in a sunny location free from shade, day in and day out, 12 months a year, for optimal performance.

As noted above, solar batch water heaters heat water during the day. In the United States and other more developed countries, however, solar batch water heaters are typically plumbed into a home's water heater. Therefore, they provide preheated water to a conventional water heater.

To understand how a solar batch water heater operates in such an installation, let's trace the flow of water, beginning in the bathroom shower, say on a hot summer day (Figure 3-5). When a hot water faucet is turned on in the shower, hot water is drawn from the conventional water heater. Replacement water flows directly into the storage water heater from the solar batch water heater. However, because solar batch water heaters often produce extremely hot water, over 160°F (71°C), installers must be sure to provide a means of reducing water temperature to prevent parboiling residents. This is achieved by placing a mixing or tempering valve between the hot water line leaving the water heater and the cold water line. This "smart valve" senses the temperature of the water leaving the tank. If it is too hot, the valve opens, permitting cold water to mix with the scalding hot water, achieving the desired temperature for showers and other domestic uses.

What happens on cooler, cloudy days when the temperature of the water in the batch heater only reaches 90°F (32°C), well below a comfortable setting for a shower? Will you end up taking a cold shower on those days? No. In such instances, water temperature is maintained by the conventional water heater. When you turn on the shower, hot water flows out of the storage water heater, whose temperature is controlled thermostatically. Replacement water comes from the solar batch collector. It is warmer than line water by 40 degrees, give or take a little. This preheated water

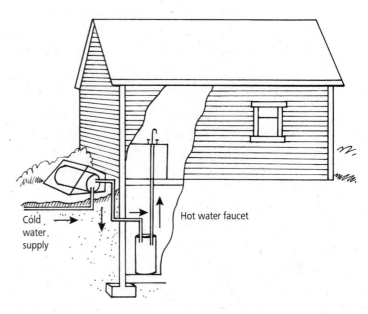

Cold water supply

Hot water faucet

Fig. 3-5: *In a solar batch hot water system, hot water for use inside the house is typically drawn from the storage water heater tank. The tank is replenished by water from the solar batch collector. Its tank is replenished by line water.*

Turn Off Your Water Heater Entirely!

During really hot sunny weather, homeowners with solar batch water heaters — and other types of solar hot water systems — often turn off their conventional water heaters. During such periods, the batch heater provides 100 percent of their needs. This is accomplished by installing a bypass valve that allows water from the batch heater to be diverted past the water heater. When a hot water faucet is turned on, water flows directly from the batch heater into the hot water line. A special mixing or tempering valve needs to be in the line, however, to prevent scalding.

flows into the conventional water heater where its temperature is boosted to the desired setting (usually around 120°F [49°C]). On such days, then, the batch heater preheats the water that flows into the storage water heater tank or the tankless water heater.

The value of this arrangement is simple: On sunny days, the batch heater provides really hot water so that your storage water heater or tankless water heater doesn't have to do a thing. It just doles out hot water as needed. On cloudy days, though, the batch heater makes the water heater's job easier by providing preheated water.

Fig. 3-6:
This modern progressive tube solar batch water heater (a) is much less obtrusive than older solar batch water heaters. It mounts on the roof (b).

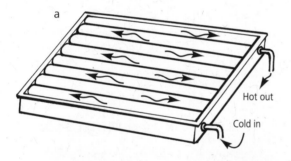

Hot out

Cold in

The water heater will operate, but it won't need to run very long to bring water up to a suitable temperature. As a result, a solar batch water heater can save energy and reduce utility bills even when not operating full tilt.

If the solar batch heater just described is too bulky for you, don't despair, there's a sleeker model that might fit the bill. It is known as a *progressive tube solar water heater*, and is shown in Figure 3-6b. This sleek four-foot by eight-foot collector mounts on the roof of a home and consists of long four-inch copper water pipes aligned horizontally (Figure 3-6a). The copper pipes are coated on the outside with a selective surface (sometimes called black chrome). Selective surface materials absorb sunlight (visible light) more efficiently than conventional black surfaces and convert it into heat. The heat is then transferred to the water inside the pipes. (Selective surfaces also reduce heat radiation back through the glass, and thus boost the efficiency of the water heater.)

The progressive tube solar water heater is glass-covered, like a batch heater, and highly insulated. The insulated metal box attaches to the roof via metal roof mounts and can withstand winds up to 180 miles per hour, according to the manufacturer.

As shown in Figure 3-6a, hot water is drawn off the top of the solar batch water heater here and is replaced by cold water that enters at the bottom of the collector. As water flows through the collector, it is heated by the sun.

Like other batch water heaters, the progressive tube solar water heater serves as a preheat tank on cooler, cloudy days, but can meet 100 percent of a family's hot water needs on warm, sunny days. Like other batch water heaters, this one requires no pumps, sensors, controls, or other moving or electronic parts. The progressive tube solar water heater comes in 30-, 40-, and 50- gallon sizes. When full, these units weigh upwards of 600 pounds — so be sure your roof can handle the load!

Pros and Cons of Batch Heaters

Batch heaters have no moving parts and are therefore the most reliable solar water heaters on the market today. They are available through solar suppliers such as Gaiam Real Goods, but can also be manufactured at home using common building materials — an old water heater tank, wood, glass, and black paint.

Batch heaters are inexpensive and relatively easy to install, although basic plumbing skills are required. Batch heaters don't require a pump, either, as noted above. Nor do they require temperature sensors. Because they're free of electronic controls, sensors, and pumps, they also require no electricity to run. They operate on line pressure — the water pressure in the pipes of your home.

Solar batch water heaters tied to domestic hot water systems provide year-round water heating in areas where freezing temperatures rarely occur. However, they'll even work in areas that experience an occasional freeze, such as northern Florida. That's because the large mass of heated water stored in the tanks resists freezing, making batch heaters immune to an occasional freeze — so long as it doesn't last too long. (In such locations, though, the supply and return lines may be susceptible to freezing, so it is wise to insulate them well and to keep pipe runs short.)

Solar batch water heaters are not without problems, however. Most notably, they're of no value in colder climates — places where freezing temperatures are a more common occurrence. You can pretty much forget about installing a solar batch water heater for year-round hot water if you live in Minnesota, North Dakota, Maine, or even Kansas — or any other area with long, cold winters. If you live in a climate that freezes with any regularity, though, you may want to install a batch heater anyway but only use it during the spring, summer, and fall. You'll need to drain the system when freezing temperatures arrive. If you want year-round solar hot water, though, you should consider one of the other systems you'll learn about shortly.

Another issue to be aware of is that while batch heaters are cheap, shipping costs can be significant. These units are pretty heavy! Don't forget to add this expense when calculating the system cost. When filled with water, batch heaters are even heavier, so roofs need to be capable of supporting the additional weight, as noted earlier. In older homes, installation of a solar batch heater may require additional supportive roof framing. When contemplating this option, check your roof framing to be sure that it is up to the task. It is best to consult a structural engineer or contact your local building department to get their advice *before* you order a solar batch water heater.

Another potential disadvantage for many families is that solar batch water heaters may require a slight change in lifestyle. To get the most from a batch heater, you will need to synchronize your family's hot water use patterns with the batch heater's hot water production. Many homeowners who've installed these systems, for example, make an initial draw of hot water early in the afternoon, for example, to run a dishwasher or clothes washer because the water inside the tank is quite hot at this time.

After drawing off hot water, water temperature in the batch heater falls. If the sun is still shining, however, the batch heater will reheat, so there's plenty of hot water by the end of the day. Showers can be taken in the early evening to utilize the hot water remaining in the tank. You can still take a shower in the morning, but remember that batch heaters cool down at night. Replacement water flowing into the storage water heater from the batch heater will be a tad cooler than it was at the end of the day. But don't worry. You won't be showering in cold water at night; shower water comes from your water heater, which maintains a comfortable temperature for your shower. The water heater will just have to work a little harder, as it is being fed slightly cooler water. To combat nighttime cooling, you can cover the glass covering of the batch heater with a thick layer of insulation each night. This will help maintain high temperature, but will involve a bit of work on your part.

Clearly, to make the most of a batch heater, homeowners need to carefully manage water use. Such systems can work well for retirees or those of us who work at home. We can reschedule clothes washing and showers more easily than those who dart off to work each day. Another downside of solar batch water heaters is that they're designed for smaller households of two to four people. But like so many things, there are ways around this limitation. If you need more hot water, you can always install two (or more) batch heaters side-by-side (in series) to boost hot water production.

Separate Collection and Storage Systems

As you have just seen, batch solar hot water systems combine storage and collection in one unit. Because of this, the water storage tank is exposed to the cooler outdoor air at night or on cold days, with obvious disadvantages. For a system that performs year-round, solar designers have separated the collection and storage functions. They have placed the collection unit, the solar panels, on the outside of the house, usually on the roof, and sequestered the storage tank inside the building, where it is much warmer. That way, hot water generated during the day isn't lost to the cold evening air. As a result, these systems can operate efficiently even in the coldest weather.

Separate collection and storage systems make up the bulk of the solar hot water systems in use today. The two most common types of collectors in use today in these systems are the flat plate collector and the evacuated tube collector. Let's take a look at each type of collector, before we explore the different types of DSHW systems.

Flat Plate Collectors

The flat plate collector consists of a glass-covered insulated box. Inside are copper pipes attached to a flat absorber plate, all painted black to optimize solar gain. (The black paint is usually a selective surface material described in the accompanying sidebar.) The pipes in a flat plate collector can be arranged in series, which means that the water flows in one end and out the other end, as shown in Figure 3-7a or parallel as shown in Figure 3-7b.

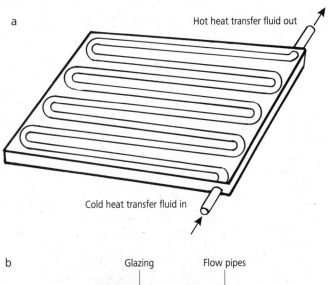

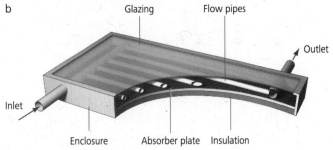

Flat plate solar collector

Fig. 3-7a and b: *Flat plate collectors are the most popular models. (a) Copper pipes in this collector form a continuous run. (b) In this more common design, water flows into the collector at the bottom, then runs upward through parallel copper tubes where it is heated.*

Sunlight entering a flat plate collector is absorbed by the dark-colored absorber plate where it is converted to heat. A heat transfer fluid flowing through the pipes absorbs much of the heat and transports it out of the collector. This liquid delivers the heat to the well-insulated solar water tank. It stores hot water that is fed into a storage water heater or tankless water heater.

Flat plate collectors are the most commonly used collector on the market today; they are useful for lower temperature applications, that is, applications that require water temperatures under 140°F (60°C), for example, domestic hot water. They're also useful in two space-heating applications: radiant floor and forced-air heating systems (discussed in Chapter 5). But they often don't produce high enough temperatures to work with baseboard hot water systems but can work in such systems with careful planning.

Evacuated Tube Collectors

Evacuated tube collectors, like those shown in Figure 3-8 a and b, are a relative newcomer on the solar hot water scene and are a serious departure from conventional flat plate collectors. These solar collectors consist of numerous (20 to 30) long, parallel glass or plastic tubes. Inside each tube is a copper pipe (absorber tube) coated with selective surface material. It runs down the center of an absorber plate, which increases the surface area for absorption. Air is pumped out of the glass or plastic tube, creating a vacuum, hence the name *evacuated tube collectors*. (Vacuums are poor conductors of heat and therefore great insulators.)

Inside each black copper pipe is a heat transfer fluid (typically methanol). It absorbs heat created when sunlight strikes the black selective surface of the absorber plate. Methanol flows upward naturally — by convection — to a heat exchanger at the top of the unit. Here, heat is transferred to another heat transfer fluid, typically a high-temperature non toxic antifreeze

Selective Surfaces

Researchers have discovered unique ways to capture solar energy and convert it to heat. One of them is a special coating applied to solar hot water panels, called *selective surfaces*. Selective surfaces have been around since the 1950s. They are coatings that are much more efficient at absorbing sunlight (visible light) than a coating of black paint. (They have a "higher absorbance.") Not only do they absorb more light and convert it to heat, they also lose less heat than a black painted material. That is, they don't re-radiate as much heat back into the surrounding environment. (They have a "lower emittance.") Combined, these two features increase the efficiency of solar collectors. Although selective surfaces cost a bit more, they're well worth it, especially in colder climates where higher collector efficiencies dramatically boost solar hot water system performance.

— HELIODYNE

(propylene glycol). It carries the heat to a solar water tank where it is transferred to water and stored for later use. Cooled methanol returns to continue the cycle.

As Bob Ramlow, solar hot water expert and senior author of *Solar Water Heating*, pointed out to me in a personal communication, there's a lot of hype about evacuated tube solar collectors so it is important to consider their pros and cons carefully. For example, the vacuum in a collector may help reduce heat loss, but it also reduces snow melt. In regions with frequent snows, these collectors shed snow poorly, especially when mounted parallel to a roof. Some collectors lose their vacuum, too, although some manufacturers have made changes in their design to prevent this.

Interestingly, numerous side-by-side comparisons have shown that evacuated tube collectors do not outperform flat plate collectors, either, despite manufacturer claims. Nor do they perform any better than flat plate collectors in "less-than-optimal regions."

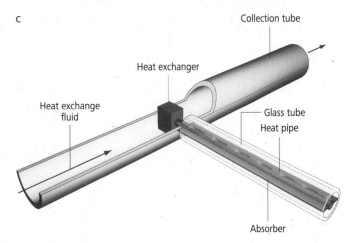

Fig. 3-8 a and b: *Evacuated tube solar hot water collectors can be mounted on the roof of a building (a) or on the ground (b). Each evacuated glass tube in this collector houses a black absorber tube filled with methanol (c). Sunlight striking the absorber tube heats the methanol, which rises inside the tube, releasing heat in the heat exchanger, and then returns to repeat the cycle.*

Evacuated tube solar collectors do produce higher temperature heat, which may work well in many baseboard hot water heating systems. As Ramlow noted, evacuated tube collectors will produce water as high as 220°F, compared to 180°F for flat plate collectors. Although this may seem advantageous, water can boil in the storage tank, a situation that typically happens when families go on vacation during the summer. Rarely will water boil in a flat plate collector system. To protect against this, you should strongly consider installing a larger storage tank with evacuated tube collectors.

"When investing in evacuated tube collectors, a buyer must pay particular attention to quality because while some of the highest rated collectors are evacuated tube type, most of the lowest rated collectors are also evacuated tubes," Ramlow noted. "Each has its place and best and worst applications. While one is not better than the other, flat plate collectors dominate all markets except China where inexpensive, government subsidized, not freeze proof glass tube ICS systems dominate."

Thermosiphon Systems

Now that you know the types of solar collectors in use today, it is time to explore solar hot water systems with separate collection and storage. We'll begin with the simplest of systems. The most basic solar

hot water system is known as a *thermosiphon system*, shown in Figure 3-9. As you can see, this system consists of a collector, a storage tank, and pipes connecting the two. Notice, however, that there are no pumps, sensors, or controls in the system. There's no need for them. Water flows through the system naturally by convection. How does this work?

Convection is the movement of a hot fluid such as air or water. As you may recall from high school physics, air and water both expand when heated. When they expand, they become less dense. This causes them to rise. In the process, they carry heat with them. But that's not the end of the story. A liquid rising by convection creates a vacuum that draws fluid in; the result is a natural pump called a *thermosiphon*. Consider an example: sunlight striking the Earth heats its surface. Heat is transferred to the air above it, causing it to expand and rise. Hot air rising, however, creates a vacuum that draws cooler air in from neighboring areas to fill the void.

In a thermosiphon solar hot water system, like that shown in Figure 3-9, water rises in the pipes inside the collectors when heated, drawing cooler fluid in from elsewhere. The result is a natural pumping action, a convective loop. The convective loop propels the hot water into a solar storage tank in the house, where the heat is deposited. Cooler water from the bottom of the tank is drawn into the collector where it is heated.

The convective flow of liquid in this system is a simple, non-mechanical pump that operates throughout the daylight hours, stripping heat from the solar collectors and depositing it in the storage tank in the house. Thermosiphon systems are simple and elegant — and less expensive than the more complicated pump-driven systems, discussed shortly. Like solar batch water heaters, thermosiphon systems are not, technically speaking, active systems. They have no mechanical pumps.

Before you race out to buy such a thermosiphon system, however, there are several things to consider. First, note that the hot water tank must be located *above* the solar collectors by about two feet. This is the only configuration that will allow natural thermosiphoning to occur. As a result, solar hot water panels are typically mounted on the ground, slightly below the hot water tank, as shown in Figure 3-9.

Thermosiphon systems can also be mounted on roofs, so long as the water tanks are located slightly above the panels, for example, in an insulated space in the attic or in an upstairs room. Some tanks are even mounted on the roof themselves, although in most applications the tanks need to be very well insulated for this to work.

Thermosiphon systems use two types of heat transfer liquid: water, or a special

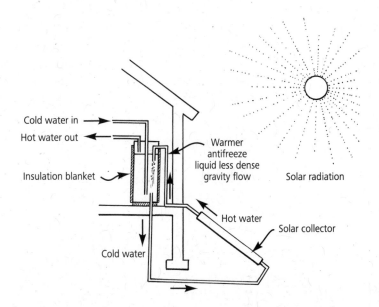

Fig. 3-9: *Convection drives the heat transfer fluid in this thermosiphon system, eliminating the need for a pump and electricity. Direct systems use water as the heat transfer fluid. Indirect systems use antifreeze as the heat transfer medium. Note the absence of a heat exchanger in this system.*

nontoxic antifreeze, the food-grade propylene glycol, mentioned earlier. Water can be used as a heat transfer liquid in systems installed on homes in warm climates where freezing is not expected. Food-grade antifreeze is used in systems installed in colder climates where freezing is expected. As a rule, water systems are simpler, cheaper, and a bit more efficient, for reasons explained shortly.

In this type of system, water heated by the panels flows directly into the solar hot water tank. It is then drawn into the storage water heater or tankless water heater and used directly any time a faucet is turned on. This configuration is referred to as a direct or open-loop system. Antifreeze systems are more complicated, a bit more costly, and slightly less efficient than open-loop water systems. Efficiency suffers a bit and costs go up a bit because the antifreeze must travel through a heat exchanger in the first tank. The heat exchanger is typically a coil of copper pipe in the wall or in the base of, alongside, or inside the tank.

As the solar-heated liquid flows through the heat exchanger, heat is transferred from the propylene glycol to the water in the tank. Cooled antifreeze flows back to the panel to be reheated continuously during the day. This type of system is referred to as an indirect or closed-loop solar hot water system. (Because the heat transfer fluid does not mix with domestic hot water, it's in a closed loop.)

Pros and Cons of Thermosiphon Systems

Open-loop thermosiphon systems are relatively simple, inexpensive, and relatively trouble-free. Closed-loop systems are a bit more complicated, in large part because of the addition of the heat exchanger and the use of propylene glycol as the heat exchange fluid. One of the problems with propylene glycol is that it begins to deteriorate at high temperatures, turning into an acidic sludge that can gum up the pipes. When this occurs, it needs to be drained and replaced, a job best reserved for a professional.

Pump Circulation Systems: Open- and Closed-Loop

Your next choice is a pump circulation system. This system is pretty similar to the thermosiphon system except that the force that moves the heat transfer liquid is a small electric pump. Most systems use AC

PV-Powered Solar Hot Water Systems

If you are concerned about the electricity required to run the pump in a solar hot water system (to be honest, it's really not that much) or if you want to go totally solar and simplify your system, you can run a solar hot water system off a small solar electric or PV module — a 30- or 50-watt PV. In such systems, a PV module is mounted next to the solar hot water panels and is wired to a DC pump. (PVs produce direct current electricity.) When the sun rises, the PV panels begin to produce DC electricity. Electricity flows to the pump, waking it up from its evening snooze. The pump begins to propel the heat transfer fluid through the collectors and pipes when they reach a critical temperature — preset by the installer so as not to circulate cold water through the system. Heat generated by the solar collectors therefore begins to flow from the panels to the storage tank. When the sun sets, the system shuts off.

PV direct systems are simple and cost-effective, and eliminate the need for sensors and controls that can break down. One manufacturer, Sol-Reliant, produces a sleek system with a built-in 35-watt PV panel. It is one of the easiest systems to install. You can learn more about it at solreliant.com.

pumps that run off household current, but, as noted in the sidebar, another very smart option is a DC pump powered by a small solar electric (PV) panel — so long as the pump doesn't have to move water too high.

Solar collectors in pump systems are usually mounted on roofs, but can also be mounted on racks on the ground, so long as they are in full sunlight all year round. Solar hot water tanks are located inside homes, usually in basements or utility rooms next to the conventional water heater that is soon to be relegated to the status of backup water heater.

The small electric pump, which is located inside, forces the heat transfer liquid up through the panels and then back down to the thermal storage tank.

Closed-Loop Water Systems: Drainback System

Pump-driven systems may use water or propylene glycol as the heat transfer fluid. As in thermosiphon systems, discussed earlier, using water as the heat transfer fluid allows one to create an open-loop or direct system. Although open-loop systems have their advantages and may, therefore, seem desirable, there is a major challenge that must be overcome in colder climates: freezing.

Freezing is a problem because when water freezes, it expands, and not just a little bit. It expands quite a lot. Water freezing inside a copper pipe can create enough force to crack it wide open. Solar designers have solved this problem in two ways. The first is a draindown system. However, this type of system is *not* recommended by most of the folks I've talked to, so I won't discuss it here. There is a viable alternative to the draindown system, a simpler trustworthy cousin, the drainback system (Figure 3-10). "Although similar in name to the draindown system," writes Ken Olson, an expert on solar hot water systems, "the drainback system is far different and much more reliable." Drainback systems rely on gravity to drain water from the pipes and panels when the circulating pump stops. When the sun stops warming the panels, a sensor signals the circulating pump to cease operation. Water flows out of the system into a storage tank. This is the type of system we installed in 2010 at The Evergreen Institute to heat water for the instructor residence and student kitchen and showers.

Gravity drainback systems are simpler than draindown systems because they have no motorized valves. This, in turn, reduces the possibility of system failure and deep freeze. Remember: in active solar systems, simplicity reigns supreme. The fewer the parts, the less likely the system is to malfunction — and the less maintenance will be required.

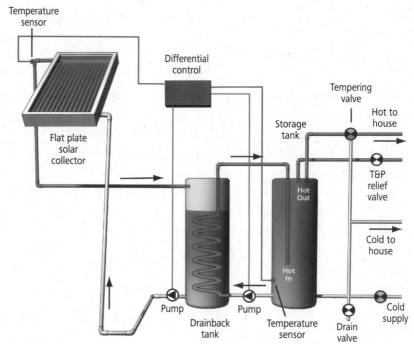

Fig. 3-10:

This is a drainback system, generally recommended by solar installers over the draindown system. The drainback tank is generally much smaller than shown here.

Gravity drainback systems can be used in all climates and are less expensive and easier to maintain than other active systems. According to Greg Pahl, author of *Natural Home Heating*, direct-circulation, gravity drainback systems are "considered by many people to be one of the simplest and best systems to install." They not only eliminate mechanical parts that can fail, they do not use propylene glycol heat exchange fluid which, as noted earlier, deteriorates over time and must be replaced every five years or so. Although a better option, these systems need to be designed and installed correctly to ensure complete drainage when the pump stops.

In some installations, like ours, only a single tank is used. Water circulates through the collector, then into and through an 80-gallon storage tank. When we turn on a hot water faucet, line water flows through 60-feet of copper pipe coiled inside the storage tank. This heat exchanger creates enough surface area so that line water is heated from 50 to 120 degrees as it flows through the pipe. (This is an on-demand solar hot water system.)

Pros and Cons of Drainback Systems

Drainback systems are simple, reliable, and less expensive than glycol-based systems, discussed next. They can be installed in cold climates and operate without propylene glycol (which comes with its own set of problems). They're fairly easy to install, too, if you know what you are doing.

On the downside, these systems require the largest pumps of all solar hot water systems in use today. Also, care must be taken when installing the system to ensure that when the system shuts down, water flows freely back into the water tank. Be sure to use distilled water in the collector loop. Never use tap water, which contains minerals that can deposit on the inside walls of the pipes and collectors, obstructing flow.

Closed-Loop Antifreeze Systems

Many active systems in use today are pump-driven systems that employ propylene glycol, a food-grade antifreeze, as the heat transfer fluid. These systems are indirect or closed-loop systems (Figure 3-11). In order to prevent the mixing of the heat transfer fluid with domestic hot water, these systems require a heat exchanger to transfer heat from the antifreeze to the water in the solar hot water tank.

Closed-loop antifreeze systems require other components as well, some of which are not required in other DSHW systems. For example, as shown in Figure 3-11, closed-looped antifreeze systems require an expansion tank. Located in the antifreeze loop, it accommodates the expansion of the antifreeze as it heats up during normal operation, preventing pressure from building to dangerous levels

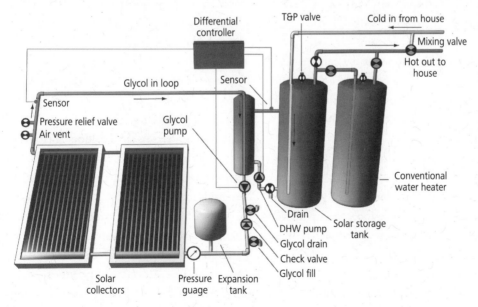

Fig. 3-11: *This system uses a small electric pump to propel antifreeze (food-grade propylene glycol) through a closed loop. Heat from the antifreeze solution is released at the heat exchanger, warming water in the solar hot water tank. The heat exchanger may be next to the tank, as shown here, under the tank, or in the tank.*

inside the glycol loop portion of the system. When the antifreeze heats up, it expands, creating internal pressure. Rather than splitting the pipes open, excess is shunted into the expansion tank, reducing pressure. When pressure drops, the antifreeze empties from the expansion tank.

Closed-loop antifreeze systems also require a drain and fill valve, a valve that allows service personnel to drain the antifreeze and then refill it, respectively.

Pros and Cons of Pump Systems with Antifreeze

Closed-loop antifreeze systems are popular, well understood by those who have been in the industry for a while, and reliable. They work well in all climates, hot or cold, and in cold climates provide excellent protection against freezing. On the downside, closed-loop systems are the most complex of all hot water systems on the market today. They have more parts and, almost without exception, the more parts there are, the more chances there are for things to go wrong.

Another small downside of this system is that its use of a heat exchanger means that it functions slightly less efficiently than an open-loop pump-driven system in which water serves as the heat exchange fluid. In addition, propylene glycol needs to be replaced by a professional from time to time. This isn't a job for most homeowners. Glycol becomes thick and sluggish if overheated, a problem that may occur as a result of pump or expansion tank failure or long power outages.

WHICH SYSTEM IS BEST FOR YOU?

Although there are a lot of options, it's really not that difficult to select a system that will work for you. Get your felt tip marker out, as Ken Olson summarizes the options: "If you live in a freeze-free climate, you should choose a batch heater or a small thermosiphon unit." Although these systems will only serve one to three people, you can always install a couple of batch heaters or thermosiphon solar collectors side-by-side to boost your hot water production. Or, says Olson, you should consider "an open-loop direct pump system circulating water from the storage tank to the flat plate collector." That's a drainback system.

In colder climates or in regions with hard water, Olson recommends "one of the closed loop systems with antifreeze and a heat exchanger." A closed-loop antifreeze system is recommended in regions with hard water because hard water contains minerals that can deposit on the inside of pipes of open-loop systems. Over time, these minerals accumulate, reducing flow rates and the efficiency of an DSHW system. The open-loop water systems just aren't worth it in such applications. Using water as a heat transfer medium in such instances, is going to cause you trouble.

SIZING YOUR SYSTEM

Sizing a DSHW system is pretty straightforward process. Before you get started, however, it is important to be sure you have a good solar site. That is, you need to be sure you can position solar collectors so that they're exposed to bright sunlight from 9 a.m. to 3 p.m. every day of the year. Fortunately, most homes, and even many apartments in cities, have good solar access somewhere on the site. Roofs are often free from obstructions that can shade solar panels.

If you have a good site, your next task is to scour your home for ways to make it more efficient with respect to hot water use. Remember: efficiency is the first rule of renewable energy system design! Make your home as efficient as possible, *then* size the system.

In his excellent article, "Solar Hot Water: A Primer," published in *Home Power* magazine Issue 84, Ken Olson recommends the following steps to make your home more efficient (these were also mentioned in Chapter 2):

1. Turn down the thermostat on your water heater to 120° to 125°F (48° to 51°C). "Many water heaters," says Olson, "are set between 140° and 180°

F (60° and 82°C)," but much lower temperatures are just fine.
2. Wrap the water heater with an insulated water heater blanket.
3. Fix drips in faucets in the kitchen and bathrooms to prevent the waste of hot water.
4. Install water-efficient showerheads and faucet aerators to reduce hot water use.
5. Insulate hot water pipes in unconditioned space.

You can also enact various conservation measures, like taking shorter showers, washing dishes by hand and not leaving the hot water running, and washing clothes in cold water.

Once water and energy efficiency and conservation measures are in place, it is time to size your solar hot water system. The size of a system depends on many factors, the most important of which are (1) your climate — how hot, cold or sunny it is; and (2) your family's water consumption. It also depends, in part, on your solar exposure. Will your system have unobstructed access to the sun from at least 9 a.m. to 3 p.m. each day? If not, you'll need a larger system.

In the United States, most families consume 15 to 30 gallons of hot water per person per day for showers, baths, washing clothes, and dishwashing. By conserving

Solar Pool Heating

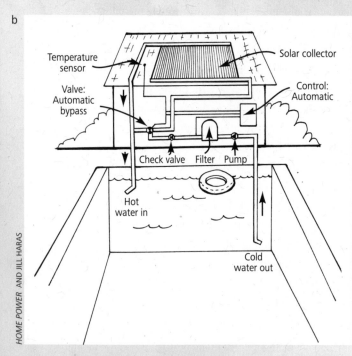

Got a swimming pool you'd like to heat with the sun? No problem. Solar hot water systems make great pool heaters. Solar pool heaters circulate water from the swimming pool to the panels and back again during daylight hours (Figure 3-12). These systems are about as simple as you can get. There's no need for a heat exchanger, antifreeze, expansion tanks, or a hot water storage tank as in many DSHW systems. Sure, you will need sensors to switch the system on and off automatically each day, and you'll need an electric pump. But you can avoid the sensors altogether by installing a PV panel that runs a DC pump, as mentioned earlier in this chapter. You'll also need a few more panels than you would if you were heating domestic water.

Solar hot water systems can warm our pools and lengthen the swim season substantially, as shown in Figure 3-13. And, if you like, you can even link the domestic solar hot water system with the pool heating system. Some homeowners have installed systems that heat their pools, their

Fig. 3-12: *(a) Swimming pools can be heated by the sun. Such systems not only provide warmer water, they can extend the length of the swimming season.*
(b) Diagram of solar pool heating system.

domestic hot water, and their homes. To learn more, I strongly urge you to read "Solar Pool Heating," Parts 1 and 2, by Tom Lane in *Home Power* magazine, issues 94 and 95.

Fig. 3-13: *This graph plots the water temperature in a swimming pool in northern Florida. As illustrated, solar pool heaters result in warmer water and also extend the length of the swimming season.*

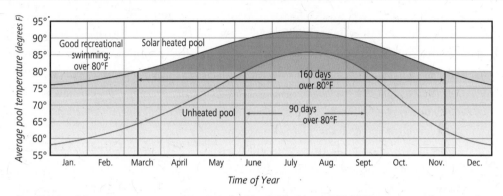

Conditions: 20-year average weather data for North Florida. 1,000 BTU per square foot per day of collector output for a pool in full sunlight; a screened-in pool would typically be 5°F lower year-round. Pool blanket used when night-time temperatures are below 60° F.

water, you can easily slash daily water consumption to the low end of the range, about 15 to 20 gallons per day. Knowing a family's daily water consumption, designers next turn their attention to daily storage capacity. If, for example, a family of three consumes 20 gallons per day per person, they'll need 60 gallons of hot water storage to meet their needs. For most households of four, Olson recommends an 80-gallon hot water tank, based on daily water use of 20 gallons per person. Once you've determined storage capacity, you turn your attention to solar collectors. You will need to determine the number of square feet of solar collectors for your application.

Once again, this process is pretty straightforward and highly dependent on solar availability and local climate. Generally, the sunnier and warmer the climate, the fewer square feet of collector you'll need. The cloudier and cooler your region, the more square feet of collector you'll need.

As a general rule, say the folks at AAA Solar in Albuquerque, New Mexico, in the sunniest locations, like the desert Southwest and Florida, you'll need about one square foot of collector for two gallons of tank capacity. Thus, for an 80-gallon tank, you'll need 40 square feet of collector. (A single 4- x 8-foot collector provides 32 square feet of collector surface.)

In the Southeast and the mountain states, which are a little less sunny and, in the case of the mountain states, also a bit cooler, you will need one square foot of collector for 1.5 gallons of tank capacity. For an 80-gallon tank, then, you will need 53 square feet of collector. In the Midwest and Atlantic states, you will need one square foot of collector per gallon of tank capacity. An 80-gallon tank will therefore require 80 square feet of collector. In New England and the Northeast, which are the least sunny areas and pretty cold in the winter, you need one square foot per 0.75 gallons of tank capacity. An 80-gallon tank will then require 107 square feet of collector.

Solar systems designed to these general guidelines will provide about 100 percent of your domestic hot water needs in the summer, and about 40 percent in the winter, says Olson. If you want to obtain more hot water in the winter, you will need a larger system. But bear in mind that you will have a huge surplus in the summer.

Designers use these general guidelines to determine how many flat plate collectors are needed for a solar hot water system. Because all solar collectors differ with respect to their efficiency, be sure to consult with a local vendor or your supplier *before* you order collectors — if you are going to install the system yourself (this is not a project for individuals with little or no experience in plumbing and building). Remember, if you undersize your system, you can always add another collector later, if you have room for one.

FINDING A COMPETENT INSTALLER

Homeowners with considerable plumbing and electrical skills can install their own solar hot water systems. You can even make your own collectors, although I don't recommend this approach for any system other than the batch heater, the simplest of all solar systems. I don't mean to thwart creativity and independence, but as a general rule it is far better to purchase a reliable, well-built system and install it yourself or hire a professional to do the work for you than to build and install your own system.

Why not build your own solar hot water system?

Solar collectors are exposed to wind, hail, and extreme temperatures. They live a very difficult life up there on our rooftops. Over the years, manufacturers have made great strides in solar collector design and materials and are producing solar collectors that can withstand the harsh treatment Mother Nature doles out. It is unlikely that you could come close to matching the solar collectors on the market today. One of the trickiest aspects of solar design is selecting black paint or coatings to paint the interior of the collector. Although store-bought black paint might work well for a few years, it may vaporize. When paint turns to vapor, the vapor leaves deposits on the inside of the collector glass, impeding solar gain and making the panel look pretty ugly.

Unless you're pretty good at plumbing and electrical work, you should probably let a professional install your system. They're fast and experienced and will be done in the blink of an eye. For the homeowner, installation might take a full weekend interspersed with numerous trips to the hardware store to pick up plumbing supplies. A professional will have everything he or she needs in the truck.

If you do decide to install your own system, I recommend you check out the excellent series of articles published in one of my favorite resources, *Home Power* magazine. Issue 94 covers installation basics — information that applies to all pump-driven systems. Issues 85 and 95 cover closed-loop antifreeze systems, and issues 86 and 97 describe the installation of drainback systems. *Home Power* continues to publish many articles each year on solar hot water systems, so be sure to check more recent issues for the latest information.

THE ECONOMICS OF DOMESTIC SOLAR HOT WATER SYSTEMS

In "Solar Hot Water Primer," published in *Home Power*, Issue 84, Ken Olson writes, "An initial investment in a solar water heating system will beat the stock market any day, any decade, risk free. Initial return on investment is on the order of 15 percent, tax free, and goes up as gas and electricity prices climb." Although I share Ken's enthusiasm for solar hot water, his endorsement is based on a comparison of a solar hot water system with a conventional electric storage water heater. Unfortunately, the economics don't always work out that well, and homeowners need to be aware that the economics of the decision depend on many factors.

Let's start by comparing a solar hot water system with an electric storage water heater. Clearly, as Olson indicates, a solar hot water system compares quite favorably with an electric storage water heater, especially if you are paying more than 8 to 10 cents per kilowatt-hour for electricity. Here's how the math works out according to Olson: a typical 80-gallon electric hot water tank serving a family of four consumes approximately 150 million BTUs over its 7-year lifetime. If your electricity costs 8 cents per kilowatt-hour, hot water will cost approximately $3,600 in US dollars over that period. At the end of its short, useful lifetime, the water heater may need replacement, further adding to the cost. (To be fair, the water heater will very likely last a bit longer than Ken estimates, and there are ways to ensure a much longer life span, as noted earlier in this chapter.)

In such instances, solar hot water systems make eminent sense. If you live in a warm climate, you can purchase and install a solar batch heater for much less than $3,600. If you live in a colder climate, you can purchase and install a pump-driven antifreeze system for $5,000 to $6,000 minus a 30% Federal tax credit in the US, bringing the cost to $3,500 to $4,200 — depending on the type of system you select (flat plate vs. an evacuated tube collector). From that point on, you will be getting hot water essentially free of charge.

The economics of a solar hot water system may be even better if you live in a state or city or are served by a utility company that offers financial incentives that offset the initial cost of the system. In many locations, homeowners can receive substantial financial incentives that dramatically reduce the initial cost of a solar hot water system. The US Government currently offers a 30 percent federal tax credit for homeowners and business owners. Business owners can

apply an accelerated depreciation schedule to a solar hot water system and can receive their tax credit immediately (within 80 days) by filing an application (grant) through the US Treasury Department. They can receive the tax credit as a grant even if they have no tax liability or their tax liability is lower than the tax credit. It's not a bad deal — if you own a business.

Solar hot water systems also make economic sense compared to propane-fired water heaters. Propane is often used in rural areas to heat homes, provide cooking fuel, and to heat water. Although propane is not as expensive as electricity, it is still generally cheaper to generate hot water from a solar system than from a propane water heater, according to Alex Wilson, Jennifer Thorne, and John Morrill, authors of one of my favorite energy books, *Consumer Guide to Home Energy Savings*.

Because natural gas currently costs much less per BTU than electricity and a bit less than propane, solar hot water systems don't always make financial sense compared to natural gas-fired hot water systems. (Although rising natural gas prices will clearly render this judgment obsolete soon.) When contemplating a DSHW system in a home served by natural gas, contact a local installer who can run the numbers for you. And, as one of Colorado's premier solar architects, Jim Logan points out, in addition to installing a system to offset high natural gas prices, "you may want to do so to offset carbon dioxide emissions" that are causing devastating climatic changes. A solar hot water system is another way to put your values in action.

In closing, installing a solar hot water system is generally a smart move, and will very likely make more sense in more places as natural gas, propane, and electricity prices continue their upward spiral. With the information you've learned in this chapter, you can now proceed with confidence. I recommend that you select a solar system design that will work in your climate, and then shop around. Contact local suppliers/installers. See what they have to offer. Read up on each of their systems *before* you lay your money down. Talk to people for whom they've installed systems. Be sure to check the Better Business Bureau for any possible complaints about the company.

Good luck!

CHAPTER 4

FREE HEAT

PASSIVE SOLAR AND HEAT PUMPS

For nearly 25 years I lived in a rather cold climate, 8,000 feet above sea level in Evergreen, Colorado, in the foothills of the Rocky Mountains. Our winters were long and cold, with evening temperatures falling well below freezing week after week. It was not unusual to have negative 20 degrees Fahrenheit for two weeks in a row. In contrast, our summers were short and cool, and spring occurred in a flash between the long cold winter and the short, cool summer. If you blinked, you might miss it. Fall passed quickly, too, morphing rapidly into winter long before many of us were willing to let it go.

Despite the chilly nature of our climate, I had no heating bill. Why?

My home was heated passively by the sun — with a cord of wood for backup heat to take the edge off those cold winter nights and to provide heat for long cloudy spells. Based on what my neighbors paid for heat, I estimate that I saved $18,000 to $20,000 in the 14 years I lived in my solar house — money I've tucked away for my retirement. You too can rack up huge savings in fuel bills by tapping into the sun's generous supply of energy. And you don't have to live in the sunbelt to take advantage of solar energy. Even if you live in a cold, cloudy region of North America — areas like Buffalo, NY or Portland, OR (places we in the solar industry refer to as the "Gloom Belt"), you can supply up to half of your annual heat requirement from the sun, perhaps even more.

How can you do this?

Through active and passive solar retrofits — the subject of this chapter and the next two chapters. If you have a good southern exposure, you can discover what the ancient Greeks and the Anasazi Indians of North America learned thousands of years ago: the sun is an amazing source of free heat.

WHAT IS PASSIVE SOLAR HEATING?

Passive solar heating is a heating system that has only one moving part: the sun. (Of course, the sun doesn't move; it's the Earth's rotation that accounts for sunrises and sunsets.) Passive solar can provide space heat for all kinds of buildings, from homes to offices to fire stations to airports to schools. My new office at The Evergreen Institute is heated passively now that I've retrofitted the building. You name the building, and passive solar can be used to warm the interior.

Passive solar design relies on ordinary visible light and near infrared radiation (heat radiation at the end of the infrared spectrum that is nearest to visible light) from the low-angled winter sun. These low-angled rays penetrate south-facing windows during the heating season. Inside buildings, the sun's rays are absorbed by floors, walls, and other solid materials. There it is converted to heat that provides warmth on the coldest of winter days.

Unlike active solar systems, passive solar design does not rely on complicated equipment or sensors. It relies on ordinary building elements, such as south-facing windows, overhangs (eaves), insulation, and airtight design. Competent designers incorporate several additional features too, for example, additional mass (concrete, stone, etc.) inside a building to absorb heat during the day. Absorbing excess daytime heat prevents a home from overheating during the day. This heat stored in mass, technically referred to as *thermal mass*, releases its stored heat at night, helping to maintain comfortable interiors day and night.

Passive solar heating relies on the fact that the sun angle from the horizon (the altitude angle) varies during the year, as shown in Figure 4-1. In the summer, the sun is high in the sky. It beats down on the roofs of our homes, so very little penetrates windows. Overhangs keep it from shining in and heating our homes in the summer. During the winter, the sun carves a low arc across the sky. The low-angled winter sun penetrates south-facing glass, warming our homes and places of work.

Passive solar heating is ideally suited to new home construction because architects can design the entire house around this simple and effective concept, coordinating all features of the design to achieve maximum year-round performance and

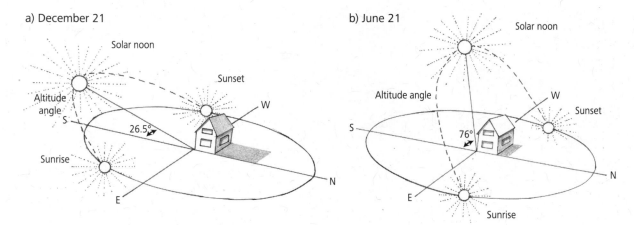

a) December 21

b) June 21

Fig. 4-1: *In a passive solar home, the low-angled winter sun penetrates south-facing windows to heat the home. During the summer, the high-angled sun shines on the roofs of houses, reducing heat gain.*

comfort. It can also be incorporated into existing homes, as I explain in this chapter and in *Home Power* magazine, Issue 138.

One of the most important aspects of passive solar design is orientation. By orienting a new home properly, so that its long axis runs east and west, a solar designer can ensure maximal solar gain. This orientation ensures you will have plenty of south-facing wall, which can be fitted with windows to allow the winter sun to enter. Performance depends on additional windows on the south side of the home to let the sun in and a reduction in north-facing windows, which can lose lots of heat in the winter. To ensure that the building stays cool in the summer, designers add overhangs to protect the south-facing glass but also reduce the number of windows on the east and west sides to reduce solar heat gain from the sun as it courses through the sky.

Performance also hinges on the inclusion of thermal mass to absorb heat, as noted above. Sealing up the cracks in the building envelope and high levels of insulation — very high levels — are also essential to effective passive solar design.

All these simple but effective measures add up. The net result is a house that can easily acquire 50 to 80 percent of its heat naturally from the sun. In some climates, like sunny New Mexico and Arizona, you can achieve even higher performance.

My concern in this book is primarily with passive solar retrofits — adding passive solar features to existing homes that have a decent solar exposure (Figure 4-2). For those who are building anew, I suggest you read my book, *The Solar House: Passive Heating and Cooling*.

Passive solar provides many benefits. One of them is that much of what you do to passively heat a building also reduces

DAN CHIRAS

Fig. 4-2: *This passive solar addition provides heat and additional living space. Although it looks nice, the two tiers of glass tend to cause overheating in the spring and fall because the lower tier of glass is not shaded. Two-story glass walls also provide an avenue for excessive heat loss in the winter. Be sure to insulate the glass at night with thermoshutters or quilted window shades.*

cooling loads (the amount of energy used to keep a home cool). Put another way, many of the design features of state-of-the-art passive solar homes help keep buildings cooler — much cooler — in the summer, dramatically cutting cooling costs.

Retrofits can also help many families reduce their electrical bills if their homes are heated with electricity. In addition, passive solar design increases natural light, known as *daylighting*, further cutting electrical demand. Thus, passive solar retrofits can help wean our society from its costly dependence on natural gas, fuel oil,

and nuclear energy. In sum, passive solar creates a win-win-win situation for everyone, except for the big energy companies. That is, it is good for people, good for your personal economy, and it is good for the environment.

IS PASSIVE SOLAR FOR YOU?

To determine if a passive solar retrofit is feasible for your home, you'll need to begin by assessing your home's solar resource — specifically, how much sunlight strikes your home during the heating season (that part of the year that requires heat). Good solar exposure is essential from October through March, April, or May depending on your location. In most cases, you will need to heat starting in the late fall or early winter into early or mid spring.

As a basic rule, your home needs to be in a location that ensures good solar exposure on its south-facing wall from around 9 a.m. in the morning to 3 p.m. in the afternoon during the heating season. This time slot is the main "window of opportunity," that is, the main period during the day for collecting sunlight during the heating season. Earlier and later hours, though they may be sunny, won't provide as much solar energy, so don't sweat if your house is shaded from sunrise to 8 or 9 a.m. or is shaded after 3 p.m. The 9-to-3 window will avail you of about 85 percent of the sun's radiation.

Good solar exposure also means that the south-facing wall (preferably the longest wall of the building) is not shaded by neighboring buildings, privacy fences, evergreen trees, or anything else. Without a good clear view of the sun, retrofitting for passive solar just won't work as well. (If your south-facing wall is shaded, but your *roof* is exposed to the winter sun, however, you may be able to heat your home with one of two active solar systems — a solar hot water or solar hot air system — described shortly.)

As just noted, an ideal home from a passive solar perspective should be oriented so that its long axis runs from east to west. This orientation allows for the largest surface area of exposed wall and window to the low-angled winter sun, which permits maximum solar gain. Determining the orientation of a building is not as easy as you might think. Here's what you do: First, take a compass to the south side of the building. Hold the compass in your hand, very level. The compass needle will point north and south. This is *magnetic* south. For proper orientation, however, you want true south. True south corresponds to the lines of longitude, which run from the North Pole to the South Pole. Interestingly, true south and magnetic south rarely line up. This phenomenon is known as *magnetic declination*. In Figure 4-3, the lines on the map indicate magnetic

Fig. 4-3: *Magnetic declination in North America.*

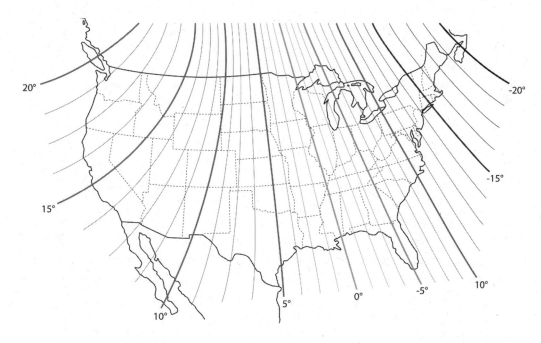

deviation. If you look in the very center of North America, say at St. Louis, you will see the line is labeled with the number 0. This indicates that, in this location, true south and magnetic south line up perfectly. As you head east or west, however, the numbers increase. This indicates that as you move away from the center of the continent, the deviation between true south and magnetic south increases. If you are in Denver, Colorado, for instance, true south is 10 degrees off from magnetic south. But which way? East or west?

In all the locations west of the center line, true south is east of magnetic south; this is called an *easterly declination*. So, in Denver, true south is 10 degrees east of magnetic south. As one travel east, say to western Pennsylvania, true south is 10 degrees *west* of magnetic south.

So, be sure to take magnetic declination into account when trying to determine the orientation of your home — or a new home you are about to build. For best solar gain in the winter, and lowest solar gain in the summer, it's best if the building is oriented exactly to true south. Deviations will decrease the solar gain in the winter, when you need it the most, and also increase solar gain in the summer, when you need it the least. Both will increase annual fuel bills.

Unfortunately, many streets in cities and towns run north and south, and many home builders, over the years, have oriented the homes they've built so that the long axis of the houses face the street — so they point east or west. Because many homes are rectangular, the south-facing walls are small compared to the east- and west-facing walls. Making matters worse, many homes are packed in suburbs like sardines in a tin, so there's not sufficient sunlight on south-facing walls to make a passive solar retrofit worthwhile.

If your house is oriented toward the street, all is not lost. You may still be able to retrofit it for solar heat, provided the south side is not shaded, for example, by a neighbor's home or fence or trees. But don't expect to heat your whole home. You may only be able to passively heat the south-facing rooms. You may only be able to acquire 10 percent or so of your heat from the sun. If your home is oriented east or west, you won't achieve huge reductions in your annual heating bill. That said, if you have dutifully sealed air leaks, super-insulated your home, replaced energy-inefficient windows or added storm windows, added insulated window shades, and taken other conservation and efficiency measures I outlined in Chapter 2, the combination of the conservation and solar could easily slash your heating bills by half. As always, don't retrofit until you have made all of the vital improvements in energy conservation and efficiency.

TYPES OF PASSIVE SOLAR DESIGN

If your home meets the criteria just described and you'd like to try passive solar, you will have three basic choices for a passive solar retrofit: direct gain, indirect gain, and isolated gain. Although that may sound mind-boggling, it's really not.

Direct Gain Retrofits

Direct gain passive solar is the most common type of passive solar design for new construction (Figure 4-4a). It has limited uses in retrofits, however. In a new home incorporating direct gain design, the long axis of the home is oriented east and west, creating the largest possible southerly surface for solar gain. As noted earlier, the designer concentrates windows on the south side to absorb the low-angled winter sun — but not too many! See the sidebar "Window Allocation for Passive Solar Homes" for recommendations on the amount of glass to use. Sunlight enters the south-facing windows and, as noted above, is absorbed by solid surfaces in the house. It is then converted to heat — or infrared radiation, as it is known by scientists. The heat warms the room. Some heat is

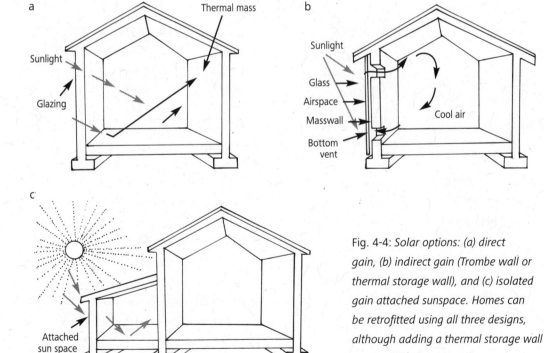

Fig. 4-4: *Solar options: (a) direct gain, (b) indirect gain (Trombe wall or thermal storage wall), and (c) isolated gain attached sunspace. Homes can be retrofitted using all three designs, although adding a thermal storage wall can be difficult and costly.*

absorbed in thermal mass inside the home for nighttime heating. In this design, heat is gained directly by the structure, hence the name, *direct gain*.

To retrofit a home for direct gain, you'll very likely need to add a few windows along the south side of the home or install larger windows to maximize solar gain. Adding windows is a job best reserved for professionals. It typically involves construction or reinforcement of rough window frames to ensure that windows won't crack and walls won't collapse. You may even need to obtain a building permit from your local building department for this work. They'll send an inspector to check out the work at various stages. A professional usually handles such details, pulling the permits and arranging for inspections. For direct gain to be successful, you need to install windows with a high solar heat gain coefficient, a topic discussed in Chapter 2.

Retrofitting for direct gain may also require you to increase the size of the overhang, or add an overhang or some other kind of shade structure, if there is none. Overhangs shade windows and walls and protect a home from overheating during the summer. In some cases, you may need to reduce the size of overhangs to optimize solar gain. A contractor can help you with this, too, although it's not going to be cheap.

Direct gain retrofitting is a great way to invite the winter sun into a home or business to provide free, natural heat and to reduce one's demand for fossil fuel energy. It does, however, involve careful planning to avoid sacrificing privacy. Also, be sure that you are not letting in so much sunlight that you decrease the utility of the rooms in your home. Bright sunlight streaming in during the daytime, for example, can render a home office where computers are used during the day almost useless. Too much light causes glare on computer screens that can lead to severe eyestrain. Even window shades won't help unless they completely block out the sun.

Solar direct gain retrofitting is tricky and presents many challenges. But there

Window Allocation for Passive Solar Homes

For new passive solar homes, south-facing glass should fall within the range of 12 to 18 percent of the total square footage of the home, depending on the climate and the amount of solar gain you desire. A 1,000-square-foot home should have 120 to 180 square feet of south-facing glass. North- and east-facing glass should not exceed 4 percent of the total square footage, and west-facing should not exceed 2 percent. These guidelines are for homes with 9-foot ceilings. If you are building a home or retrofitting a home with vaulted ceilings, you need to take into account this fact. A room with an 18-foot ceiling will require twice as much solar glazing (south-facing glass) as a room with a 9-foot ceiling.

are ways to make the job easier. Rather than retrofit an existing wall, you can build an addition that incorporates direct gain passive solar design features, as shown in Figure 4-2. The addition contributes heat to adjoining rooms. If you are thinking about adding a room or two to your existing home, and the room faces south, why not heat it passively — for free?

No matter what you do, be sure that solar glazing (south-facing glass) in retrofits or additions can be insulated at night — either by rigid foam insulation internal shutters, as shown in Figure 4-5, or by a quilted fabric like those used to make Warm Windows insulated shades, available at fabric stores. Good window insulation dramatically improves the performance of a passive solar home! Remember, the sun only shines into south-facing windows, at most, 8 hours a day in the dead of winter. Windows are a weak point in the thermal envelope of a home. Most of the day, heat flows out of windows. So, be sure to stop that flow by installing and using insulated shades.

When retrofitting a home for passive solar, you'll want to be certain that you don't end up overheating it. Overheating can be prevented by several measures, for example, by installing overhangs. A two- to three-foot overhang works in many climates. Overheating can also be avoided by installing window shades and by installing additional thermal mass. (Thermal mass usually consists of masonry materials such as concrete, tile, or bricks that absorb sunlight and hold heat.) You can boost thermal mass in an existing home by removing wall-to-wall carpeting to expose wood floors. (Although wood is not as good as masonry material, it does provide thermal mass in a home.) Floors exposed to direct sunlight can be tiled to improve the performance of your retrofit.

When retrofitting your home for passive solar, it is important to proceed carefully. Work out a plan before you start adding new windows and think through

Fig. 4-5: *Thermoshutters like these are made of rigid foam insulation and plywood, covered with a decorative fabric. Closing the thermoshutters at night or on cold, cloudy days greatly reduces heat loss.*

KEN WOODS

the repercussions of each change you want to make. Pay special attention to thermal mass. If the total square footage of the solar glazing will exceed 7 percent of the floor space, you'll very likely need to add thermal mass. If not, the incidental mass, that is, the thermal mass in drywall, framing, floors, cabinets, and furniture, usually suffices to absorb heat and prevent overheating. To determine how much thermal mass your home will need, you may want to contact a local passive solar designer/architect or hire a solar consultant. Or, you can pick up a copy of my book, *The Solar House,* for guidance on the subject, too.

Indirect Gain Retrofits: Thermal Storage Walls

Indirect gain passive solar is so named because it involves an intermediary structure that absorbs the sun's energy, converts it to heat, and then transfers the heat to the house (Figure 4-4b). That intermediary is a Trombe wall (pronounced "trom") named after a French engineer who invented this innovative passive solar design feature. Trombe walls are also called *thermal storage walls,* a term I like to use.

Thermal storage walls are built from solid materials such as poured concrete, cement blocks, bricks, adobe blocks, or even rammed earth (the latter two are natural building materials). These materials store considerable amounts of heat. Thermal storage walls are located along the south side of buildings. The outside of the wall is fitted with glass — usually double-paned glass mounted three to six inches away from the thermal storage wall. Low-angled winter sun penetrates the glass and is absorbed by the surface of the mass wall. It is then converted into heat that warms the surface of the wall. The heat then migrates slowly into the mass wall, eventually reaching the interior surface from which it radiates into the adjoining room.

The thickness of the mass walls is specified so that the heat conducting through the wall during the day reaches the interior of the wall early in the evening, around sunset. The wall will continue to radiate heat into the room throughout the night. For daytime heating, builders often install a window or two in the wall. This allows for direct gain, and permits views of the outside.

Daytime heating can also be achieved by placing vented openings in the thermal storage wall, as shown in Figure 4-4b. These openings allow room air to circulate through the space between the wall and the glass, drawing off heat.

Thermal storage walls work well in many climates, but because they are so massive, they're not an option for retrofitting most homes. That's because most foundations in existing homes are not designed to support the weight of the mass wall. If you want to retrofit your home to include this effective passive solar design, you'll very likely have to beef up the foundation, a procedure that is not only tricky, but costly.

So why mention the thermal storage wall at all?

If you are building a new home or building an addition, a thermal storage wall is a wise choice. It works well in a variety of climates, from warm to extremely cold. For details, you may want to take a look at the section on thermal storage walls in my book *The Solar House*.

Isolated Gain Retrofits: Attached Sunspaces

The final option for passive solar design is the isolated gain system, more commonly referred to as the *attached sunspace* or *solar greenhouse* (Figure 4-4c). Attached sunspaces are passive solar heat collectors built onto the side of buildings. They are heated by the sun; the heat they generate is then transferred to adjoining rooms. Hence the term, *isolated gain*. (Heat is gained in an isolated space.)

Attached sunspaces are relatively easy to build onto many homes, provided there's adequate solar exposure. All-glass attached sunspaces are available in kits, and there is no shortage of installers who

can put one in for you. Unfortunately, this design is fraught with problems.

All-glass designs — that is, attached sunspaces with glass walls and glass roofs — tend to overheat in the summer and fall, causing severe discomfort in the home. They may even overheat in the winter. But isn't this structure designed to collect heat and transfer it to the house in the winter?

Absolutely, but don't expect to be able to use the space for much else. Overheating renders the sunspace much too hot to enjoy during daylight hours in the winter. Moreover, all but the hardiest of plants (cacti and succulents) find the intense heat oppressive. Most plants need to be kept below 85°F (29°C) for optimal growth; photosynthesis grinds to a halt at 100°F (37°C).

Despite the fancy brochures and advertisements you receive in the mail touting the value of all-glass attached sunspaces, you'll very likely be disappointed if you go this route. I recommend that you think long and hard about the downsides of this retrofit. You're basically buying a solar oven. To prevent overheating, you'll need to cover the glass much of the year in most climate zones. The results can be quite hideous, as Figure 4-6 demonstrates.

A more effective attached sunspace design is shown in Figure 4-7. This design has a solid roof, which permits solar gain during the late fall, winter, and early spring,

DAN CHIRAS

Fig. 4-6: *Attached sunspaces are an easy way to retrofit a home for passive solar, although they often don't perform well. This all-glass attached sunspace must be covered during the late spring, summer, and fall to prevent overheating. All-glass attached sunspaces also tend to lose lots of heat at night and therefore must be isolated from living spaces by doors.*

when the sun is in an intermediate and low position in the southern sky. Sunlight penetrating the south-facing windows warms the interior of the attached sunspace; the warm air created in the sunspace can then be transferred to adjacent rooms. During the summer, the solid roof all but eliminates unwanted solar gain.

While attached sunspaces of this nature are the best option in new and existing homes, you still need to design them very carefully to ensure optimal performance, especially in terms of maximum heat transfer into the house; otherwise, they'll

basically heat themselves and provide very little, if any, additional heat to your home.

Here's what you need to do to make an attached sunspace work optimally: First, install openings in the wall between the sunspace and the adjoining rooms to permit warm air to flow into the house. To promote passive air movement, install openings near the floor and the ceiling of the wall between the sunspace and the adjoining room. This helps to create a convection current that circulates warm air into the house. (Note that a door opening into the space rarely suffices for heat transfer.)

To improve heat transfer from the sunspace to the house, I recommend installing a thermostatically controlled fan in the window or in an opening between the sunspace and the house. (Be sure to install a quiet one. Look for a low-sone fan with a rating of around 1.0.) DC fans can be powered by a 20- to 50-watt photovoltaic module. When struck by sunlight, the module generates DC electricity, powering the fan. When the sun sets or is covered with clouds, the fan shuts off. You won't need any fancy switches or controllers of any sort with this sort of system.

Second, be sure that the openings can be closed off at night to prevent heat from escaping into the cooler sunspace.

Third, be sure the wall between the attached sunspace and your home is insulated to prevent heat inside the house from escaping at night into the cold interior of the sunspace. You may also want to consider building a mass wall between the two, as I did on a home office I retrofitted for passive solar (Figure 4-7).

Fourth, be sure the ceiling and foundation of the attached sunspace are well insulated. Ceiling insulation should be in the range of R-50 to R-60 in most areas. To make the floor, I'd recommend pouring a four- to six-inch concrete slab or excavating and backfilling with crushed rock. Insulate around the perimeter of the foundation and under the slab to retain heat. Use rigid foam insulation rated for underground applications. Two to four inches of blueboard or pink board (extruded polystyrene) will generally suffice. Fifth, for best performance, use low-e double-paned glass (described in Chapter 2). Be sure the glass has a high solar heat gain coefficient, which means that it permits lots of sunlight to enter the structure. Ratings of around 0.5 or higher are ideal for many locations. In southern regions, like Florida, solar heat gain coefficient should be much lower, around 0.35 or so. A knowledgeable glass supplier will know what you mean when you tell him this and should be able to help you select the best glass. If he or she doesn't know what you are talking about, call someone else. You'd be amazed at

Fig. 4-7:
This mass wall in the attached sunspace I designed for my previous home heats up during the day and radiates heat into the adjoining room at night.

DAN CHIRAS

how many window installers don't understand their product.

Sixth, for the absolute best performance, insulate the glass at night. I recommend that you use rigid foam insulation panels placed against the glass between the framing members to reduce heat loss. Or you may want to install insulated shades. (Of the two, rigid foam panels are better; they have a higher R-value.) Rigid foam insulation panels will require additional work on your part each day; you'll have to take them out in the morning and put them back at night. Raising and lowering shades will also require some extra effort, but it's less work than inserting foam insulation panels. Either way, the additional labor on your part is a small price to pay to keep the interior of an attached sunspace warm at night. The sunspace will also warm up much more quickly the next day, resulting in more heat gain in your home — where you want the heat!

Seventh, be sure to install some operable windows in the attached sunspace to bleed off hot air during the summer and fall. To ventilate naturally, that is, without fans, you'll want to create a convection current by installing a couple of opening windows low in the structure and a couple opening windows up high — or include a roof vent. Because warm air rises, it will escape through the upper windows or roof vent, drawing cooler air in through the lower windows. You can purchase openers that operate automatically — opening a vent when the temperature reaches a certain level.

As noted above, attached sunspaces are the "easy option" for solar retrofits. They're pretty easy to integrate into an existing home and relatively easy to build. They're not very expensive, either, and they go up quickly, providing instant savings and comfort.

Attached sunspaces also provide a space for a solar cooker like the one shown in Figure 4-8. A solar cooker is a remarkably simple device consisting of a box with reflectors that concentrate the sunlight striking the device, a dark interior to convert sunlight to heat, a glass lid to hold in the heat, and a shelf to hold cooking pots or cookie sheets. You can purchase one online through Gaiam Real Goods or make your own. Numerous plans to build your own are available online at solarcooking.org/plans.htm.

Attached sunspaces of the type I'm recommending generally perform much better than the all-glass variety. They stay cooler in the summer, provide more usable space, and provide adequate levels of heat during the winter. Still, they do have some downsides.

For one, they make lousy growing spaces. If you are hoping to grow vegetables year round in an attached sunspace with a roof, forget it. You'll be disappointed by the results. That's because many vegetables do best when the sunlight comes from above. A lack of summer sunlight often

Fig. 4-8:
Solar cookers like this one can be used to bake cookies or bread, or make a variety of one-pot meals. You can build your own cooker inexpensively by using aluminum foil, glass, and a cardboard or wooden box.

DAN CHIRAS

causes plants like tomatoes to become tall and thin — we say they become "spindly." Spinach and lettuce like overhead light, too. Although such plants may thrive in the sunspace during the winter, especially if you can keep the temperature up at night via the insulation strategies I've mentioned, they'll languish in the shade inside the sunspace in the summer, because the sun is high in the sky, beating down on the roof and overhangs. As a result, very little sunlight will penetrate the structure. To offset this problem, you can install a few standard skylights in the roof. Unfortunately, this could lead to summertime overheating. In addition, skylights lose huge amounts of heat on cold winter nights, causing the space to chill down, thwarting plant growth. Another option for providing overhead light in the summer is the solar tube skylight, shown in Figure 4-9. Solar tube skylights consist of a

Fig. 4-9: *(a) Tubular skylights allow light into a home, thus reducing daytime lighting and electrical demand. They lose much less heat than conventional skylights at night. (b) Hallway before installing tubular skylight. (c) Hallway after installing tubular skylight.*

small glass or plastic dome-shaped lens mounted on the roof that collects sun and directs it into a polished aluminum tube that extends from the roof to the ceiling. (The polished aluminum ensures maximum light transmission.) Light enters the room through a ceiling fixture, a diffuser that disperses light.

Solar tube skylights let in lots of light from a rather small opening — much smaller than standard skylights; as a result, they minimize unwanted heat gain caused by conventional skylights. This in turn considerably reduces the threat of summertime overheating. Because they utilize a small opening, tubular skylights also minimize wintertime heat loss at night, a huge problem with conventional skylights.

Another problem that you may find with an attached sunspace is that it won't be habitable during the winter, except perhaps during the morning before the sun has warmed it. Once the sun begins beating in, however, you'll fry.

Like other forms of solar retrofitting, you will have to obtain a permit from the building department for an attached sunspace. They'll inspect the project at various stages and upon completion of the project.

GETTING THE HELP YOU NEED

Passive solar retrofits, combined with energy efficiency modifications, can help propel you and your loved ones toward energy independence. Although I've outlined many of the ideas and methods to achieve this laudable goal, be sure to research these options in more detail. My books, *Solar Home Heating Basics*, *The Solar House*, and *Green Home Improvement* provide a wealth of additional information on energy efficiency and solar heating.

Also, don't be afraid to ask for help. Although it may cost you a bit, the assistance of a skilled and knowledgeable solar architect or builder can help you avoid costly mistakes. Don't assume that just any architect can help you, however. Even ones who claim to be interested in doing more solar may not have the experience you need.

I strongly recommend that you work with an architect who can perform a computerized energy analysis of your home to determine how the changes you and he are proposing will affect energy consumption. Running an analysis on your home with the proposed efficiency and solar retrofits in mind will not only give you an idea of how the changes will improve the performance of your home, they may enable you to discover more ways to save on fuel and increase comfort. Bear in mind, though, that some of the more sophisticated energy analyses, like Energy-10, can be costly.

To reiterate, whenever possible hire an architect or builder who has experience in passive solar design. I recommend working with designers and architects who actually live in passive solar homes, although, regrettably, not many that I've encountered practice what they preach.

Why hire someone who lives in a solar house?

Architects and designers who live in solar homes often develop a strong appreciation of the art of building comfortable, energy-efficient structures. Without this day-to-day experience, it's my belief that many designers miss key points. They may, for example, fail to include sun-free zones — areas where family members can relax, work, or watch TV without being blasted by bright sun. (Sun-drenching has rendered many a new passive solar home unlivable on bright, sunny winter days.)

Bear in mind, too, that one design does not fit all. A passive solar design or a passive solar retrofit that works in Minnesota might overheat in Kansas or Tennessee. You and your architect need to design specifically for your region using computer software that allows you to assess the performance of a design — before you build it. Your job in retrofitting your home is to do the most for the least, but don't cut corners. Use high-quality windows, for instance, and insulate, insulate, insulate. Before you insulate, however, seal up the many air leaks in the building envelope of your home or business. They're robbing you blind!

SOME FINAL THOUGHTS ON PASSIVE SOLAR RETROFITS

Passive solar retrofits offer many advantages, as I've noted. For additions, direct gain systems are really quite economical. If you and your architect are smart, you can design and build a direct gain passive solar addition for little more than a conventional addition would cost. You will be blessed with totally free heat immediately.

Retrofitting an existing home or business will always cost money, however. For direct gain systems, you'll need to tear out and replace existing windows. For isolated gain systems, you'll need to build an attached sunspace. Unfortunately, I've never seen any figures on the economics of various retrofits that determine costs and savings or determine the return on these investments. That leaves the task up to you. My suggestion is to compute the costs of the retrofit, then estimate your savings on fuel bills, being sure to take into account the rising cost of fuel. Divide the savings by the cost and you'll have the return on investment. If a retrofit costs $2,000 and will save you $200 a year on heating and cooling costs, your return on investment is $200/$2,000 or 10%. This may help to guide your decision. Bear in mind, however, when calculating the costs and benefits of a passive solar retrofit, that not only will increasing fuel costs make the economics more favorable, you will also benefit aesthetically and increase your home's value. For example, adding windows will provide better views, create a roomier feel to your home, and provide natural daylighting. Of course, passive solar retrofits will also make your home warmer and more comfortable. In addition, a passive solar retrofit could add to the curb appeal of your home, making it more desirable should you decide to sell it. An attractive, energy-efficient passive solar home with much lower fuel costs than similar homes is likely to be much more attractive to potential buyers, especially if fuel prices continue to escalate. The resale value of your home could increase dramatically as a result of your lower fuel bills.

HEAT PUMPS

Another renewable energy option that you may want to consider is a heat pump. A heat pump is an ingenious device designed to extract heat from the ground or the air around a home in the winter, concentrate it, and then transfer the heat into the interior of the structure. Heat pumps can be used as the primary heat source for new or existing homes. They're ideal for sites that aren't conducive to solar energy retrofits.

What makes heat pumps so special is that they don't burn fossil fuel like many conventional home heating systems. They operate entirely on electricity. (Electricity that may be generated from fossil fuels,

however, and usually is.) Moreover, heat pumps can be run in reverse during the summer to *extract* heat from our homes. Heat pumps fit into two basic categories: air-source and ground-source. Both are discussed below.

Ground-Source Heat Pumps

Ground-source heat pumps (GSHPs), as shown in Figure 4-10, extract heat from the Earth around a home and transfer the heat into the house in the winter, providing space heat. Ground-source heat pumps consist of three parts: (1) pipes buried in the ground to draw heat from the Earth, (2) the heat pump, and (3) a means of distributing heat in a house (a conventional heating system such as a radiant floor or forced air).

In the winter, ground-source heat pumps gather heat from the subsoil, well beneath the frost line where temperatures remain about 50°F (10°C) year round. Heat is collected from this massive heat sink by water or propylene glycol pumped through the underground network of pipes. The heat is then concentrated by the heat pump and transferred into a home. How do heat pumps turn 50°F ground heat into 80° or 90°F (27° or 32°C) space heat?

Ground-source heat pumps rely on refrigeration technology — refrigerants, gases, compressors, and pumps. Without

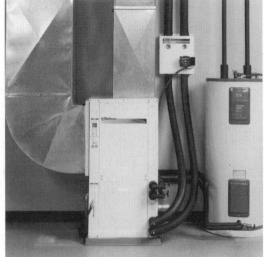

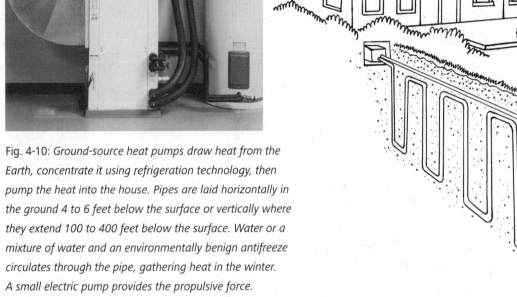

Fig. 4-10: *Ground-source heat pumps draw heat from the Earth, concentrate it using refrigeration technology, then pump the heat into the house. Pipes are laid horizontally in the ground 4 to 6 feet below the surface or vertically where they extend 100 to 400 feet below the surface. Water or a mixture of water and an environmentally benign antifreeze circulates through the pipe, gathering heat in the winter. A small electric pump provides the propulsive force.*

getting too technical, the heat delivered to the heat pump causes the refrigerant in the unit to vaporize and expand. The refrigerant is then sent to a compressor. Compressing the vaporized refrigerant releases heat. This heat is captured and transferred to the heating system, and the refrigerant gas is recompressed and reused. Heat pumps can also be used to draw heat from the groundwater or even surface waters, for example, a nearby lake or pond. These units are known as water-source heat pumps.

Pros and Cons of Ground-Source Heat Pumps

Ground-source heat pumps have many great features. Perhaps the most important is that they are extremely energy efficient. According to the United States Department of Energy (DOE) and the Environmental Protection Agency (EPA), ground-source heat pumps are the most efficient, environmentally benign, and cost-effective space heating and cooling system on the market today. These systems use relatively small amounts of electricity to power their pumps and compressors — about 25 to 50 percent less than conventional heating and cooling systems. Moreover, ground-source heat pumps require no additional fuel other than the heat they extract from the Earth, which is free. Because of this, ground-source heat pumps offer the lowest carbon dioxide emissions of any *conventional* heating and cooling system on the market today.

Yet another advantage of ground-source heat pumps is that they can be installed in virtually any climate. Although they're more expensive to install than conventional heating and cooling systems, efficiency gains pay for the additional costs in two to ten years. In addition, ground-source heat pumps carrying the EPA's Energy Star label can be financed with special Energy Star loans from banks and other financial institutions. Some of these loans offer a lower interest rate than you'd be able to get for a conventional heating system. Others allow longer repayment periods. Some combine both features. (For information on Energy Star loans, call 1-888-STAR-YES).

Another advantage of ground-source heat pumps is that they are more compact than conventional heating and air conditioning systems. They also have relatively few moving parts and typically require less maintenance than conventional heating and cooling systems. Underground piping is often warranted for 25 to 50 years.

Yet another advantage of ground-source heat pumps is that they are much less likely to set your home on fire, as they contain no flames. The absence of combustion also eliminates indoor air pollution. And, as if that's not enough, ground-source heat pumps operate fairly quietly. Further adding to the list of advantages, residential ground-source heat pumps can be fitted with a device that transfers waste heat from the compressor pump to a storage water heater. In the summer, while the unit is cooling the house, waste heat from the ground-source heat pump provides 100 percent of a home's hot water; in the winter, it provides about 50 percent.

Ground-source heat pumps do have a few disadvantages, however. The main problem is that they use a refrigerant known as hydrofluorocarbon- 22 or HCFC-22. This chemical is less stable than ozone-depleting CFCs and, therefore, tends to break up in the lower atmosphere, so it does reach the ozone layer where it can destroy ozone molecules. As most readers will know, ozone provides a protective shield against ultraviolet-B radiation, which causes cataracts and cancer, and injures plants. Although HCFCs destroy

far fewer ozone molecules than CFCs —
5,000 per molecule of HCFC compared
to 100,000 per molecule of CFC —
ozone loss is still significant. Fortunately,
ground-source heat pumps come with
factory-sealed refrigeration systems that,
according to manufacturers, will seldom or
never have to be recharged. This reduces
leak potential and ozone destruction.

And then there's the issue of cost. As
noted earlier, ground-source heat pumps
cost more than conventional heating and
cooling systems, largely due to the fact
that they require extensive excavation,
although bore holes can be drilled vertically to eliminate this problem.

Be sure to call your local utility to see
if they offer any rebates or other incentives
for installing a heat pump.

Air-Source Heat Pumps

Air-source heat pumps operate the same
way as ground-source heat pumps, but
capture heat from the air rather than the
ground. In the summer, they operate in
reverse. That is, they strip heat from the
house and dump it outside. As illustrated
in Figure 4-11, the air-source heat pump
extracts heat from outside coils filled with
refrigerant. The refrigerant absorbs heat
from the air, even at very low temperatures.

How?

The refrigerant is cold, sometimes as
cold as 0°F (18°C). Because it is cooler

than ambient air, the refrigerant can
absorb heat from it. Figure 4-11 illustrates
how this device operates.

Although sales of air-source heat
pumps outstrip sales of ground-source
heat pumps, air-source heat pumps are
not as efficient. Nor can they be installed
in as many places. As a rule, air-source
heat pumps work best in warmer climates,
such as that of the southeastern United
States, where the cooling load exceeds the
heating load. However, there are a few
cold temperature air-source heat pumps
on the market now that work in temperatures as low as 10°F (−12°C). Of the two
choices, however, a ground-source heat
pump is probably a better choice. They use
half as much HCFC as the air-source heat
pumps and deliver more heat (or cooling)
per unit of electricity consumed. If you'd
like to learn more about heat pumps, I
strongly recommend John Lynch's article,
"Heat from the Earth" in *Home Power*,
Issue 98.

CONCLUSION

In closing, although heating oil and natural
gas are heading toward extinction, we have
many options to heat our homes that make
sense economically and environmentally.
Passive solar is great for new and existing homes, as are heat pumps. In the next
chapter, I'll tackle active heating systems:
solar hot air and solar hot water systems.

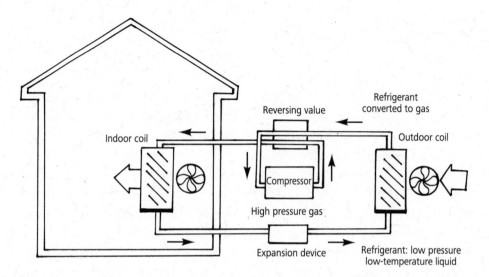

Fig. 4-11: *An air-source heat pump consists of an indoor and an outdoor portion. After giving off its heat inside a house, the cool refrigerant passes from the house to the outside. Here, the pressurized liquid enters an expansion device and is converted to a low-pressure, low-temperature liquid, which then enters the outside coil. A fan blows outside air over the cold coils. Heat is transferred from the outside air into the coils where it warms the refrigerant, causing it to transform from liquid to a gas. As the gas expands, it absorbs heat. Next, the heated refrigerant gas passes through a compressor, reducing its volume. The heated, high-pressure refrigerant then enters the house. Because it is hotter than the inside air, it gives off heat in the inside coil. A fan blows across the coils, stripping the heat away. When cooling a house in the summer, the process is identical, except the heat is obtained from inside the house and transferred to the outside.*

CHAPTER 5

SOLAR HOT AIR AND HOT WATER SYSTEMS

AFFORDABLE HEAT FROM THE SUN

Two additional options for clean, affordable space heat are solar hot air systems and solar hot water systems. They're both great for retrofitting a home and when designing a new home as well, although passive solar is ideal and is typically a cheaper choice for brand new homes. We'll examine both of these systems in this chapter, starting with solar hot air systems.

SOLAR HOT AIR HEATING SYSTEMS

Solar hot air systems capture sunlight energy striking a collector mounted on or near a home or business. The solar hot air collector heats indoor air circulating through the device, then sends it back into the building. These systems can provide years of inexpensive, worry-free comfort.

Solar hot air systems are primarily installed on homes, but they can also be used to heat offices, workshops, garages, and barns. Very large systems can be installed on warehouses, factories, and other commercial buildings.

Like other renewable energy technologies, solar hot air systems free homeowners from worries over rising fuel costs. Solar energy costs the same today as it did when humans first appeared on savannahs of Africa — nothing. Combined with other renewable energy technologies, solar hot air systems can help all nations stretch declining supplies of home heating fuels, notably natural gas and propane, giving them much-needed time to fully develop clean, affordable, and reliable renewable energy resources. Solar hot air systems

DAN CHIRAS

Fig. 5-1: *It's not that difficult to build a solar hot air collector. This fairly crude but effective system is used to heat a home in Colorado Springs, Colorado.*

could also help homeowners do their part in curbing the global climate change — a disaster now unfolding before our eyes — by lowering their carbon footprint.

In this chapter, you will learn how solar hot air systems work and how they are installed. We'll also explore costs and return on your investment.

How Do They Work?

Solar hot air systems are fairly simple devices, far simpler and easier to use than passive solar (covered in Chapter 4) or solar hot water space-heating systems (covered later in this chapter). They're also a lot cheaper, and do not require complicated electronic controls. If you're skilled, you can even build a solar hot air system yourself (Figure 5-1).

Solar hot air systems also produce heat earlier and later in the day than solar hot water systems. As a result, "they may produce more usable energy over a heating season than a liquid system of the same size," according to the DOE's online publication, "Consumer Guide to Energy Efficiency and Renewable Energy." Moreover, air systems do not freeze. Minor leaks in the collector or distribution ducts that can cause significant problems in hot water systems are of lesser consequence in hot air systems. So how does a solar hot air system work?

All solar hot air systems rely on a collector. The collector is typically mounted on the roof of a home, but may also be mounted on vertical south-facing walls or even on the ground (if it's not shaded during the heating season). As you shall soon see, vertical south-facing walls are the best location.

Solar hot air collectors capture sunlight energy and use that energy to heat room air that is circulated through them on sunny days during the heating season (Figure 5-2). The solar-heated air is transferred into the interior of the building, thanks to a small AC or DC fan or blower.

Unlike solar hot water systems, solar hot air systems are controlled by relatively simple electronics. A temperature sensor mounted inside the panel monitors collector temperature. When it climbs to 110°F, the sensor sends a signal to a thermostat mounted inside the home. It, in turn, sends a signal to the fan or blower, turning it on if room temperature is below the desired temperature setting. When the temperature inside the collector drops to 90°F, the fan switches off.

Solar hot air systems provide daytime heat on cold, sunny days, unlike solar hot water systems, which are designed for daytime and nighttime heat thanks to their ability to store heat in water tanks located inside the home. Some early systems stored hot air in rock storage bins for use at night or during cloudy periods. Although ingenious, this approach generally proved to be disappointing — quite disappointing. Designers found it difficult to achieve adequate and predictable air flow through rock beds.

Rock storage also poses a health risk. As green building expert Alex Wilson notes, rock beds "can become an incubator for mold and other biological contaminants — causing indoor air quality problems in the home." Because of these problems, most rock bed storage systems have been abandoned.

Even though solar hot air systems used today principally provide daytime heat, the heat they supply to a building can accumulate indoors during the day. For example, it may be absorbed by drywall, tile, and framing. At night, the heat stored in these forms of thermal mass radiates into the rooms during the early evening, helping

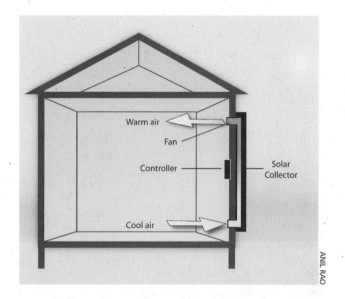

ANIL RAO

promote evening comfort. Obviously, the more thermal mass, the greater the nighttime benefit. Even so, solar hot air systems are still primarily considered daytime heat sources.

Types of Solar Hot Air Collectors

Although you may not have heard about solar hot air systems, they're not a new technology by any means. They've been around since the 1950s. Today's systems fall into two categories: open and closed loop.

Open-Loop Solar Hot Air Collectors

The newest solar hot air system is the open-loop design (Figure 5-3). Hot air collectors in open-loop systems extract cold air not from the building, but from

Fig. 5-2:
This diagram shows how air is circulated through and heated in a solar hot air collector.

Warm air

Fan

Controller

Solar Collector

Cool air

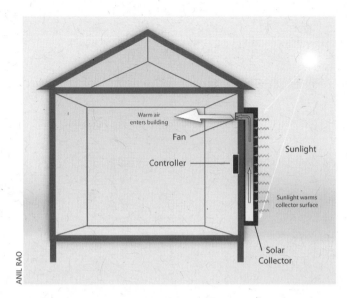

ANIL RAO

Fig. 5-3: *In an open-loop design, cold outside air is drawn into the collector where it is heated. The warmed air is then blown into the building. These systems are primarily used in large warehouses that need a modest amount of heat and are not generally very airtight.*

DAN CHIRAS

Fig. 5-4: *Absorber plate of a solar hot air collector.*

the out-of-doors. They heat this air, and then transfer the heated air into the building. Collectors used in open-loop systems are known as *transpired air collectors.*

A transpired air collector consists of a dark-colored, perforated metal facing, known as the *absorber plate* (Figure 5-4). The sides and back of the solar hot air collector are made from metal and are typically insulated to reduce heat loss. Sunlight striking the absorber plate of a transpired collector heats the surface. Heated air is drawn into the panel by a blower and is then piped to the interior of the building.

Closed Loop Solar Hot Air Collectors

A closed-loop solar hot air system, the most popular option today, consists of a glazed flat-plate collector (Figure 5-5). Inside the collector is an absorber plate that's typically roughened to increase turbulence inside the collector. (Turbulence helps strip heat from the absorber plate.) The collector is insulated and covered with single- or double-pane glass (glazing).

In closed-loop systems, cool air is drawn into the collector from the interior of the building and then heated as it passes through the collector. The heated air is then blown back into the building. Room air is typically drawn into the panel through a short section of pipe that runs through the wall or the roof of a building, although longer duct runs are also used in some applications.

Air entering the intake duct passes through the panel, either behind the dark-colored metal collector surface (back-pass collectors) or in front of it (front-pass collectors). Back-pass models come with single-pane glass; front-pass models require two-pane glass to reduce heat loss. Heated air is blown· into the building through a short length of pipe or duct, if possible. Small registers are mounted on the air intake and outtake openings that penetrate the exterior walls of the building.

Of the two types of glazed solar hot air collectors, back-pass collectors are most common. They're cheaper to manufacture and about 50 pounds lighter than front-pass collectors ·and, therefore, they are a little easier to install.

Like transpired air collectors, glazed collectors in closed-loop systems are thermostatically controlled. Both closed-loop and open-loop systems employ backdraft dampers to prevent air from flowing through the system in reverse at night by convection, a natural phenomenon that will suck heat out of a building.

Installing a Solar Hot Air System

Installing a solar hot air panel is not a job for novices. Although, as solar hot water and hot air system expert Chuck Marken notes in an article on solar hot air system

Fig. 5-5: *This glazed collector from Your Solar Home is mounted on my garage in Colorado. It contains its own source of electricity, a small solar panel that powers the fan that moves air through the collector. In 2010, my garage was insulated and converted to a candle-making facility where my long-time partner, Linda Stuart, runs her soy candle business, Evergreen Candleworks. The solar hot air panel helps heat the business.*

in *Home Power* magazine, "a seasoned crew of two can install a solar hot air system in a few hours [or even] a full day. … If this is your first time, plan on a weekend, even with help."

Mounting Options

As noted previously, the best place to mount a solar hot air panel is on the south side of a home, provided it is unshaded during the heating season. Mounting the panel vertically on a south-facing wall ensures maximum solar gain during the

YOUR SOLAR HOME

Fig. 5-6:
Mounting a
solar hot air
collector on the
south side of a
building is ideal,
as explained in
the text.

coldest months — when you need the heat the most (Figure 5-6). (The collector is more closely aligned to the incoming solar radiation from the low-angled winter sun.) Mounting on a south-facing wall also shades the solar hot air collector during the summer from the high-angled summer sun, especially if the building has adequate overhang.

The second most desirable location for a solar hot air system, though infrequently chosen according to Your Solar Home's Todd Kirkpatrick, is on a rack mounted on the ground. This is known as a *ground-mounted solar hot air collector*. The rack should be anchored to a suitable concrete foundation, for example, concrete piers.

Ground-mounted systems require considerable ductwork — as do roof-mounted systems, discussed next. Both of these systems require much heftier fans to

ensure adequate air flow. (This is due to the additional ductwork.)

Because they function primarily during the heating season, when the sun is low in the sky, solar hot air collectors should be mounted more vertically than solar electric modules, which are mounted to absorb as much sunlight as possible throughout the year. The tilt angle of the solar hot air collector, that is, the angle between the back of the module and a line running horizontally from the bottom of the collector, should be set at latitude plus 15 degrees. If you live at 45 degrees north latitude, for instance, you should mount a collector at 60 degrees. If you live at 35 degrees north latitude, the tilt angle should be 50 degrees.

One of the most popular of all places to mount a solar hot air collector, though one of the most problematic, is on the roof (Figure 5-7). Roof are popular because they are often unshaded during the heating season. That's because roofs often tower above other buildings — even mature trees in some cases — providing good access to the winter sun in densely populated urban and suburban environments.

Unfortunately, roof mounts can be more complicated and more costly than wall or ground mounts. In homes with attics, for instance, installation requires the use of flexible insulated ducts to transport air to and from the collector. Outdoors, the ducts are protected from the elements by galvanized pipe. Unfortunately, flexible insulated ducts are ribbed, which greatly increases air turbulence, which reduces air flow. Partially because of this problem, much larger fans are required for roof-mounted systems. In homes with vaulted ceilings, collectors require much shorter duct runs. The same goes for wall-mounted collectors.

Another problem with roof mounts is that they are exposed to sunlight year round, though most of us need heat only during the late fall, winter, and early spring. Intense sunlight during the summer can, over the years, damage a roof-mounted solar hot air collector, so professionals generally recommend against such installations. If that's the only location you have, though, it is probably wise to cover the collector during the cooling season.

Pointers for Mounting a Collector

Precise directions for installing a solar hot air collector are beyond the scope of this book; however, a few comments and suggestions seem in order — so you know what you are getting into.

To install a glazed solar hot air collector, you'll need to cut two large (5- to 7-inch) holes in the wall or roof and ceiling. When cutting holes in a wall or roof, be certain not to damage water pipes or cut or damage electrical wires. Work slowly,

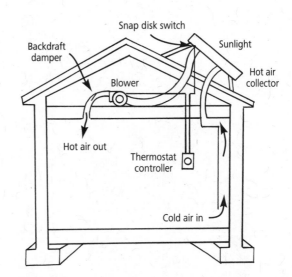

checking for potential obstructions. Cut a hole in the drywall or siding, then check for wires and pipes. Manufacturers of the collectors provide metal mounts to attach to the exterior wall. The solar hot air collectors are attached to these mounts.

Because collectors are fairly heavy and rather large, measuring around 4 x 7 feet, you'll need a couple of brawny assistants to make sure the job goes right and to protect your back and toes. (You don't want to drop one of these collectors on your feet!)

When installing a collector, you'll need to do some wiring; for example, you'll need to run electric wire to power the fan or blower. You will also need to wire the thermostat that comes with the unit to the temperature sensor mounted in the collector. Wiring diagrams can be difficult to understand for the electrically illiterate.

Fig. 5-7: *This diagram shows a solar hot air collector mounted on the roof — not an ideal location, because it receives much more heat in the summer than in the winter.*

Fig. 5-8: *This is Cansolair's solar hot air collector. See text for a description.*

CANSOLAIR

Fig. 5-9: *Cansolair's fan/filter unit conveniently plugs into a wall socket, eliminating complicated wiring.*

CANSOLAIR

To make your life easier, two manufacturers have provided some rather ingenious and simple wiring alternatives. Canada's Your Solar Home, for instance, manufactures a solar hot air collector known as the SolarSheat (Figure 5-5). It comes with its own supply of electricity: a small solar electric module mounted above the solar hot air collector. The solar electric module generates direct current (DC) electricity when struck by sunlight. The DC electricity it generates powers the SolarSheat's DC fan. All the installer or homeowner needs to do to wire the system is to connect the two wires that run the panel to the thermostat inside the house. It's about as simple as you can get, as I learned when I installed mine.

Another ingenious solution to simplify wiring is provided by Cansolair. Cansolair's Solar Max is a glazed solar hot air collector made from 240 empty aluminum cans (Figure 5-8). The cans are painted black and arranged in 15 vertical columns. The cans are housed in an attractive collector. Air flows through the solar-heated cans inside the collector. Air flow is supplied by an indoor fan located in an attractive console that plugs into a 120-volt wall outlet. It propels room air through the duct system and then through the collector and back into the building. The fan console contains a washable filter that helps to remove large particles from the air (Figure 5-9).

Although solar hot air collectors are often attached to existing walls or roofs, DeSoto Solar sells collectors that can be integrated into walls, reducing their profile. Your Solar Home has also integrated some of its panels in the exterior walls of new homes. Such installations are best reserved for brand new construction or when building an addition on a house. It is much easier to install a panel in the wall in new construction.

How Well do they Work?

Solar hot air panels are capable of boosting the temperature of air flowing through them — often quite substantially. According to the DOE, "Air entering a (glazed) collector at 70°F (21.1°C) is typically warmed an additional 70°–90° (39°–50°C)."

Transpired air collectors may provide considerably less heat than glazed collectors. Solar energy expert Chuck Marken notes that transpired air collectors only increase the temperature of the air flowing through them around 11°F (−11°C), which is probably of little value to residential structures, although useful in factories, warehouses, and even indoor lumber yards where a little bit of warming dramatically improves working conditions for employees.

Ralf Seip, a homeowner who installed a SunAire glazed solar hot air collector to heat his basement workshop in Michigan, found that his collectors raised the temperature of the air flowing through them slightly less than 70°F (40°C), but only for an hour and a half to two hours a day during the peak of solar gain on sunny winter days. (That is when solar irradiance is at its highest.) His system raised room temperature in his basement workshop by about 3.6°F (2°C) on cloudy days and 11°F (6°C) on sunny days. Although daytime temperatures in the workshop only reached 63°F (17°C), that was suitable for working.

Individuals who've installed solar hot air systems on their homes report impressive results. Steve Andrews, a residential energy expert based in Colorado, for example, installed a collector to heat the bottom 500 square feet of a his tri-level home in sunny Denver. This area was usually 5° to 6°F colder than the rest of the house. He found that the solar hot air collector "made a difference during sunny winter days and the following evenings." Although the system was of little or no help on very cloudy or snowy days, "overall, the comfort improvement was dramatic."

The SolarSheat 1500G I tested when researching an article on solar hot air collectors for *Mother Earth News*, (from which this chapter was adapted), consistently raised the temperature of indoor air entering the collector at around 68°F (20°C) by

40°F (4.4°C) on sunny cold winter days. It didn't make much difference in the room temperature of my home, but I was dumping the heat into a very large open space of over 2,400 square feet. I'm certain the collector would have made a substantial contribution in a smaller room. That leads us to an important question: How much space can a solar hot air collector heat?

Solar expert Chuck Marken recommends one 4 x 8 foot collector per 500 to 1,000 square feet of heated space, depending on the solar resources at one's location and the energy efficiency of the building. A 2,000-square-foot home, for instance, may require two to four collectors. Separate collectors may be required for each room. For larger rooms, it may be necessary to install two or more collectors, although a more powerful fan will be required to ensure adequate air flow. If collectors will be shaded during part of the day, for example, by tree limbs, more collectors will be needed. Whatever you do, be sure to seal up the leaks in your home and beef up the insulation first.

Does Solar Hot Air Make Cents?

But does a solar hot air system save you money?

Bill Hurrle of Bay Area Home Performance, a company in Wisconsin that installs solar hot air systems, notes that in their cold, cloudy climate, the best they're able to achieve when retrofitting homes is a 25 to 35 percent annual reduction in heat bills.

Although this may not sound impressive, it could cut fuel bills by $200 to $270 per year. Greater savings may be achieved in sunnier climates. As a rule, active solar heating systems are most cost-effective in cold climates with good solar resources. But they also make sense in other locations as well. Many solar installers will help you determine if a solar hot air system makes sense. Or, you can run the numbers yourself.

One popular way of determining the cost effectiveness of a solar system is to calculate their return on investment. Return on investment is determined by dividing the annual savings by the cost of the system. If a $2,100 system saves $300 per year in heating bills, the return on investment is $300/$2100 or 14.3%. Manufacturers estimate returns on investment of 12.5% to 25%, based on current energy prices.

To accurately calculate return on investment, you should take into account the rising cost of fuel and any interest you pay on money borrowed to purchase the system or interest lost if you withdraw money from a savings account to purchase a system.

To be even more precise, maintenance costs should also be added to calculations of return on investment. Fortunately, very little maintenance is required on a solar hot air system, as they only have two moving parts, a fan or blower and a backdraft damper. Fans or blowers may need repair or replacement, but not for 18 to 25 years. Backdraft dampers can also be counted on for many years of trouble-free service. The rest of the system should last 50 years or longer!

When calculating return on investment, however, don't forget to check into financial incentives from state and local government and local utilities. At this time, no federal incentives are offered for solar hot air systems unless they are dual-use systems — that is, systems that use some of the hot air to heat water for domestic uses. To avail yourself of these credits, however, the solar hot air panel must be SRCC (Solar Rating and Certification Corporation) certified.

Open-Loop or Closed?

Of the two types of systems, I personally like the closed-loop systems — collectors with glazed panels — for residential and most business applications. If you are heating a warehouse, barn, or less tightly sealed structure that doesn't need much heat, you may want to consider a transpired solar hot air system. They are manufactured by a company called SolarWall.

Transpired collectors for residential structures have been pulled from the market for a number of reasons. As Steve Andrews notes, "Transpired air collectors appear to be very suitable for a range of commercial applications, but seem to present more challenges than opportunities in residential applications — either in existing or new-home applications." One problem in residential applications was that transpired collectors introduced too much fresh air into homes.

While fresh air is required for wintertime comfort, these systems can pump thousands of cubic feet of air into homes every hour, resulting in frequent air changes — far more than is required. Air pumped into a home forces warm indoor air out through openings in the building envelope. The subsequent loss of heat through cracks and other openings in the building envelope could waste a substantial portion of the heat produced by the collector or your heating system.

Moisture buildup in walls is another potential problem of this design. Open-loop systems draw outdoor air into a house, bringing with it a lot of moisture. Because the incoming air must have somewhere to go, these systems force the now moist indoor air through cracks around doors and windows, light switches, electrical outlets, and other openings in the building envelope. The moisture may deposit in the insulation in wall cavities or in the attic or ceiling.

Moisture that collects in the wall insulation, greatly reduces its effectiveness.

As I have noted earlier in the book, even a tiny amount of moisture can decrease insulation's R-value (its ability to retard heat flow) by half.

Moisture can also promote mold growth and can, over time, cause wood to rot. Decaying wood framing members may eventually collapse, resulting in structural damage in the walls of wood-framed houses.

Although transpired collectors may not be suitable for homes in colder climates, they may work well in barns,

garages, workshops, chicken coops, or even pigeon lofts — and other less airtight structures that require only a slight temperature boost to enhance comfort of its occupants.

Transpired solar hot air collectors are great for commercial buildings, such as warehouses and shipping facilities, even indoor lumber yards.

Shopper's Guide

Solar hot air collectors can be ordered online or purchased through a growing list of solar suppliers — companies that also install other solar systems such as solar hot water or solar electric systems. When shopping, beware!: "Marketing departments can make anything look good," says Bill Hurrle. Watch out for too-good-to-be-true claims. "One collector won't heat a home," says Hurrle, despite what some sales people may tell you.

SOLAR HOT WATER HEATING SYSTEMS

Many people are aware of solar hot water systems that heat water for domestic uses — such as showers, baths, and cleaning dishes — a topic covered in depth in Chapter 3. As pointed out in that chapter, domestic solar hot water systems are economical, highly effective in many areas, and cheaper over the long haul than conventional water heating systems in most

Fig. 5-10: *This building (the faculty residence at The Evergreen Institute) is equipped with a solar hot water system for domestic hot water — for showers, baths, dishwashing, and the like. These systems are highly effective and typically designed to supply about 40 to 80 percent of a family's hot water.*

DAN CHIRAS

instances. In fact, domestic solar hot water systems represent one of the best buys among *all* the solar technologies (Figure 5-10). The only solar technology that's more economical is passive solar heating, discussed in Chapter 4.

Solar hot water systems, often referred to as *solar thermal systems*, can also be designed to heat homes and businesses. Such applications require more solar collectors and a much larger solar storage tank to hold heat for extended cold periods. Solar hot water space-heating systems also require a modification of the existing heating systems so that solar heat can be delivered through the hot air distribution system, either ducts or pipes.

As natural gas prices climb, the cost of home heating is expected to rise dramatically. Those who retrofit their homes with solar thermal systems that provide domestic hot water *and* space heat stand to save a sizeable amount of money. They'll also be able to substantially reduce their environmental impact.

In this section, I focus on solar hot water heating systems — concentrating on those that have proven to be most reliable and cost effective. If you haven't read Chapter 3 on domestic solar hot water systems, you may want to do so now. It provides the background material for this chapter. Once you understand how domestic solar hot water systems work,

you will be better able to understand the more complicated solar hot water space-heating systems. This section will help you understand your options and help you when you talk with potential installers. More in-depth coverage can be found in other books, such as Bob Ramlow and Benjamin Nusz's newly revised *Solar Water Heating*.

Bear in mind, as you study this material, just as in all other solar heating options, sealing the leaks in the building envelope, beefing up the insulation, and other efficiency measures such as water-efficient showerheads are among the very first steps you should take in your quest for an economical and environmentally sustainable solar hot water heating system.

SOLAR HOT WATER SPACE-HEATING SYSTEMS

Solar hot water heating systems are larger and a bit more complex than domestic solar hot water systems. While a domestic solar hot water system — designed just to provide hot water for showers and the like — typically contains two to four collectors, a system designed for space heating typically contains eight or more solar collectors. These systems also employ a much larger storage tank — instead of 90 gallons, they require 500- to 1,000-gallon storage tanks. The larger tanks are needed to store huge quantities of solar-heated water for use at night or during extended cloudy periods to heat a building.

Solar hot water space-heating systems are active systems — pump circulation systems, either *drainback* or *glycol-based systems*. Heat is stored in water tanks, as just noted, or in beds of sand. Let's start with water storage systems, the most common type.

Drainback Solar Hot Water Heating Systems

In a drainback solar hot water space-heating system, the design includes one large storage tank or a single drainback tank and multiple storage tanks to hold hot water for times of need. Most systems have one large, extremely well-insulated tank — typically 500 to 1,000 gallons.

On sunny winter days, the water in the storage tank circulates through the collectors, becoming hotter and hotter. However, when clouds obscure the sun or the sun sets, the system shuts down. All the water in the collectors and pipes drains back into the storage tank. If the home requires space heat, hot water inside the massive storage tank is there to provide it. To understand how heat is stripped out of hot water inside the storage tank, consider an example: Imagine a solar hot water space-heating system in a home previously heated by forced-air

gas furnace. Gas furnaces heat air, using natural gas, propane, or oil. The hot air is circulated through supply ducts in the house. Cold air returns to the furnace to be reheated.

When a solar hot water system is added to a home such as this, heat is drawn principally from the solar storage tank. The furnace serves as backup. When the thermostat calls for heat, solar-heated water flows out of the storage tank through a heat exchanger installed in the furnace. The furnace fan turns on and blows air over the heat exchanger. The air blown over the surface of the heat exchanger strips the solar heat from it. The solar heat removed from the heat exchanger is then distributed through the house via the existing heat ducts in the house. The tank continues to provide heat, circulating hot water through the heat exchanger in the furnace, so long as it is needed and so long as there's enough hot water in the tank to heat the home. If the water temperature inside the tank drops below the desired setting, the gas furnace turns on, taking over, providing backup to the solar system.

Drainback systems work well in these applications, supplying space heat and heating water for showers and the like. However, in such cases, a separate heat exchanger needs to be installed to provide heat to the storage water heater or tankless water heater.

Glycol-Based Systems

Glycol-based systems tend to be the system of choice for space heating, especially in cold climates. In these systems, solar heat can be stored in a single large tank or multiple storage tanks or in beds of sand from which the heat is later extracted and delivered to the home's heating distribution system — either through ducts or pipes.

The design of a closed-loop glycol-based space-heating system is identical to those described for domestic solar hot water in Chapter 3. However, most systems store heat in a single large, extremely well-insulated water tank located in the basement. Heat drawn from the tank is distributed through forced air, radiant floor, or baseboard hot water systems. Heat is drawn out of the storage tank via a heat exchanger loop. In these systems, a second heat exchanger inside the storage tank can feed hot water to the domestic hot water system. That is, hot water can flow from the storage tank to the storage water heater or to a tankless water heater. A separate storage water tank can also be installed for the domestic hot water, but such configurations are more expensive and much more complicated than single-tank storage that feeds into the home heating system and the domestic hot water system.

Pumps in glycol-based systems can be run by household current or by electricity

DAN CHIRAS

produced by a PV module mounted near the hot water collectors (Figure 5-11). This is known as *direct* PV, because the photovoltaic module is wired directly to the pump that circulates glycol through the collectors and back to the storage tank. In most pump circulation systems, temperature sensors and the differential controller regulate the solar loop, the circulation of heat transfer fluid through the solar collectors. In direct PV systems, the PV modules replace the differential controllers.

Solar hot water space-heating systems typically require a dump load — a place to dump excess heat during the summer. Heat can be diverted to pipes buried around the perimeter of a home. This heat may seep into a home during the winter, keeping it warmer.

Fig. 5-11: *This PV module powers the circulation pumps in the solar hot water system, not the fans or pumps required to distribute heat through a home's heating system.*

A Word on Storage Tanks

There's a lot more to solar hot water systems than is presented here, but I leave these details to you and your installer. However, a few words on storage tanks are in order.

It should be noted that storage tanks for solar hot water space-heating systems should be well-built and durable. They must be of materials capable of storing water for long periods at 180°F. High-temperature fiberglass tanks work well. Stainless steel dairy tanks work well, too. But beware of homemade storage tanks made from steel. Wooden framed tanks lined with plastic or rubber roofing materials (EPDM) may be inexpensive to build, but don't last very long. Ordinary storage water heaters aren't such a good idea either. They don't last long when storing really high-temperature water.

Ideally, tanks should be seamless and jointless to reduce the possibility of leakage. They also have to be sized so they can be carried into a home without cutting out a wall! Larger tanks made from fiberglass come in pieces that are assembled inside the building.

For years of leak-free performance, it is recommended that pipes enter and exit at the top of the tank, rather than from fittings located along the bottom. If you want to learn more about tanks or any other component of solar hot water systems, I highly recommend Bob Ramlow and Ben Nusz's book, *Solar Water Heating*. It has an excellent chapter on solar space heat. Tom Lane's book, *Solar Hot Water Systems,* is another excellent (although a much higher level) book. It is great for people who are thinking about getting into the business. For more on tanks, you can also consult with installers in your area.

As noted earlier, it is possible to store solar heat in sand beds. Hot water is circulated through pipes embedded in insulated sand beds, often situated under the concrete slab of buildings. These systems work very well — if they are designed and installed correctly. To learn more about this heat storage system, you should consider purchasing a copy of Ramlow and Nusz's book, *Solar Water Heating*.

Another option for heat storage, though, is to pump heat into the concrete slab of a building. The heat is delivered directly from the collectors into a set of pipes embedded in the slab during construction. These systems provide radiant floor heat and work well in extremely cold climates. Be sure to insulate under the slab and around the perimeter of the foundation. To prevent overheating during the summer, you may need to divert heat to a dump load — for example, to pipes buried in the ground outside the home or a large insulated storage tank buried in the ground under or alongside the house.

Choosing the System to Meet Your Needs

Despite the many options, it's not difficult to select a solar hot water heating system that will work for you. If you live in a freeze-free climate, you should consider a drainback system — an open-loop, pump circulation system. Remember that open-loop drainback systems are not generally recommended in regions with hard water, because minerals in hard water can deposit on the inside of the pipes in the solar loop. Over time, mineral deposits reduce flow rates and the efficiency of a system. In colder climates or in regions with hard water, your best bet is a closed-loop glycol-based system although drainback systems work well in these locations as well.

Sizing and Pricing a Solar Hot Water Heating System

Before sizing a system, it is important to make your home or business as efficient as possible, as noted earlier. How large a system you will need depends on many factors, including wintertime temperatures, the amount of sunshine you receive, how energy efficient (airtight and well insulated) your home is, and temperature requirements (do you require the heat set at 80°F to stay warm?).

Your best bet is to call local installers who will assess your heating requirements, the energy efficiency of your home or business, solar availability, and other factors, then recommend a system size. A small home-sized system for an efficient home in a sunny climate could cost as little as $12,000; a larger system required for an energy-inefficient home with poorer solar resources could cost many times more — $50,000, or more.

The economics of the investment, however, are generally quite favorable — that is, these systems generally represent a good economic investment. The return on investment depends in part on the type of heating system you are currently using. If you are replacing electric heat with solar heat, the investment is almost always quite sound — as it is extremely costly to heat with electricity. Replacing propane heat with solar heat can also be quite cost effective. Replacing natural gas with solar heat may have a lower return on investment, although natural gas costs have increased dramatically in the past ten years and are expected to continue to rise.

Don't forget that there are economic incentives from the federal government in the United States for solar hot water systems. You will receive a 30 percent federal tax credit for installing a solar hot water system.

A local installer can give you the rundown on state, local, and federal incentives, if any. Or, you can check dsireusa.org to see what's available in your area.

CONCLUSION

Solar hot air and solar hot water systems can reduce heating bills and improve home comfort — if properly designed and integrated into a new or existing home. Shop carefully. I strongly recommend hiring an experienced professional to install your system. Call solar hot water and solar electric installers in your area to see if they sell and install these systems. Research their products, ask to see some installations, and talk to their customers.

Before purchasing either system, be sure to investigate local building codes and zoning ordinances. You may need a building permit. Also, check out neighborhood or subdivision covenants as well. They may prohibit solar systems, although many homeowners have successfully challenged their homeowner's and neighborhood associations.

If a solar hot air or hot water system makes sense for your home or your business, you will be rewarded many times over. Once you pay off your investment, you'll receive free heat for the life of the system. At that point, you can sit back and enjoy the sun's free heat and the savings, knowing your work is done. You won't be cutting and hauling firewood to save money on your monthly fuel bill. Nor will you have to worry about rising fuel costs that are plaguing your neighbors! And, you'll be doing something positive to create a clean, healthy future.

CHAPTER 6

WOOD HEAT

M any people heat their homes with wood. With the impending decline of two vital home heating fuels, natural gas and oil, wood is very likely to become even more popular, especially in rural areas. Many urban and suburban homeowners may find wood to be an economical source of primary or secondary heat — backing up conventional fuels or renewable home heat sources such as those discussed in the last two chapters. Although this may seem like an outlandish claim, you'd be amazed at how much wood is available in and around cities from tree removal, packing crates, and discarded pallets. You may even be able to grow some of your own wood for heat if your lot is big enough and you plant fast-growing trees such as cottonwoods.

This chapter offers some guidance on wood heat, a source I've used as both a backup and a primary heat source for decades. In this chapter, I will outline several options and provide information that will help you make sound economic and environmental choices.

RETROFITTING FIREPLACES FOR EFFICIENCY

Many homes in North America have fireplaces, and many homeowners may be inclined to use them to provide heat. Before you stoke up the fire, though, consider these facts. First, fireplaces are one of the least efficient heating technologies humans ever invented. Most fireplaces achieve efficiencies of only 10 to 20 percent. In other words, only 10 to 20 percent of the heat generated by burning wood in a fireplace actually makes its way into adjoining rooms. Most of the heat is lost up the chimney. In fact, some fireplaces lose more heat than they generate! Second, fireplaces are sources of enormous heat loss in the winter, even when they are not operating. Moreover, cool indoor air leaks out of fireplaces in the summer, raising cooling costs.

Unless your home is equipped with an extremely efficient fireplace, close up your fireplace or install a fireplace insert in the opening (Figure 6-1). Fireplace inserts are steel boxes that fit into fireplace openings. Most models are about 70 percent efficient.

Fireplace inserts cost as much as wood stoves. Models vary considerably in their heat output, ranging from 30,000 BTUs per hour to nearly 85,000 BTUs per

Fig. 6-1:
Fireplace inserts are wood stoves that fit into fireplaces, dramatically increasing their efficiency.

MORGAN, INC

hour. The difference is related to their size and construction.

When purchasing a fireplace insert, be sure to select a model that comes with a blower fan. Fans circulate room air around the combustion chamber of the insert, then force it out into the adjoining room, boosting the efficiency of the stove. For more information, you may want to consult John Gulland's piece, "Woodstove Buyer's Guide," published in one of my favorite magazines, *Mother Earth News*.

FUEL-EFFICIENT WOOD-BURNING STOVES

Free-standing wood stoves are another option for home heating. Wood stoves are widely available and come in a wide variety of shapes, sizes, colors, and materials. Let's begin by looking at the three basic types: radiant, circulating, and combustion.

Radiant Wood Stoves

Radiant wood stoves are constructed of a single layer of metal — either sheet metal, cast iron, or welded steel (Figure 6-2). To protect the metal from heat damage and prolong the life of the stove, manufacturers typically line the combustion chamber with fire brick — a high-temperature brick that protects the metal.

Radiant stoves are so named because they warm rooms primarily via radiation: heat energy produced by the fire radiates off the hot metal surface of the stove. Heat is also stripped from the hot stove and circulated throughout the room by convection. Convection currents are created by the heat released from a wood stove. It warms room air near the stove. The hot air then rises. Cooler room air flows in to fill the gap. As the cooler room air flows near the stove, it is heated. The heated air rises, creating a convection loop that circulates heat through the room. Because heat may accumulate near the ceiling, some homeowners install ceiling fans to blow heat down, assisting natural convection.

Radiant stoves constitute the bulk of the wood stove market because they are simpler than the next major type, the circulating stove, and use less material. Their sparing use of material makes them less expensive to manufacture and hence less expensive to buy. Some cast iron radiant stoves are an exception; they can be quite costly due to more extravagant design (Figure 6-3).

Besides being fairly economical, radiant wood stoves are fairly efficient — with combustion efficiencies in the range of 70 to 80 percent.

When shopping for a radiant wood stove, look for models that allow you to control the flow of air into the combustion chamber. Controlling the amount of air entering the combustion chamber allows the operator to regulate the rate of

Fig. 6-2: *Radiant wood stoves like this model from Travis Industries are made from cast iron or welded steel.*

Fig. 6-3: *This cast iron radiant wood stove from CFM Majestic is attractive and efficient.*

Fig. 6-4: *Circulating wood stoves feature double-wall construction. They're a bit more expensive than radiant wood stoves.*

Creosote and Other Considerations

Many stove operators like to turn their wood stoves down once they've gotten sufficiently hot. Unfortunately, reducing the flow of air into the combustion chamber reduces the combustion efficiency of the stove. Restricting the air supply may cause a fire to smolder and produce more air pollution, especially particulates and carbon monoxide. In addition, volatile gases from the wood escape up the chimney unburned. These organic chemicals are often deposited on the walls of the flue pipe, forming dreaded creosote. If not periodically removed from the flue pipe, creosote may catch on fire, burning at a searing 2100°F (1150°C). These extremely hot fires can spread to the house.

combustion. The more air that's allowed into the combustion chamber, the hotter the fire. Because hotter fires also burn out more quickly and often generate more heat than is necessary at any one moment, air flow controls allow the homeowner to regulate heat output. Many stove users restrict the air flow to the combustion chamber once the fire has been started and is burning strong. This ensures a good, long, steady burn and helps prevent a room from overheating. Be careful, though, not to turn the air flow down too far. This can cause inefficient burning and other problems described in the accompanying sidebar.

As you shop, you will find that some wood stoves come with automatic controls. They allow the operator to set the desired room temperature; the stove then regulates air flow into the fire to maintain it. Vermont Castings' Encore wood stove is an example. This stove is also designed for smokeless top loading and has a removable ash pan for ease of cleaning — convenience features many homeowners find appealing.

Circulating Wood Stoves

The second type of wood stove is the circulating wood stove (Figure 6-4). Circulating wood stoves look like ordinary wood stoves. However, upon closer examination you will notice one striking difference that affects both price and performance. Circulating stoves are double-walled. The combustion chamber is constructed of cast iron or welded steel and is typically lined with fire brick. The outer shell is typically made from a lightweight sheet metal. Separating the two is a small air space. When the fire burns inside the stove, heat is transferred to the inner shell. This heat is removed by room air that circulates through the air space either passively (by convection) or actively (by a fan). The hot air is then vented into the room. Heat also radiates off the outer shell, but the double-walled construction prevents the outer metal layer from getting as hot as the surface of a radiant stove. This is important for safety. Thus, although circulating stoves cost more, they provide a slightly greater measure of safety, which is especially important for families with young children who might be inclined to touch a hot stove out of curiosity or who might stumble and fall against the stove. (All wood stoves should be "fenced off" from young children.)

Like radiant wood stoves, circulating stoves achieve efficiencies in the range of 70 to 80 percent, depending on the design.

Combustion Wood Stoves

Last, and certainly least, is the combustion wood stove. The old-fashioned Ben Franklin stove is a good example. Combustion stoves

Where's the Heat?

You may be surprised to know that half to two thirds of the fuel value of wood is locked up in gases and volatile liquids. When a piece of wood is burned, for example, only 30 to 50 percent of the heat comes from combustion of the solid woody fibers. The rest comes from the combustion of hydrocarbons, if your stove burns hot enough to ignite them.

From *The Solar House: Passive Solar Heating and Cooling,* by the author.

are radiant wood stoves with one significant difference: the doors can be opened when the fire is burning. Opening the doors converts the combustion wood stove into a fireplace slightly more efficient than a standard fireplace.

Open doors provide a view of the fire, which many find desirable, but they allow a huge amount of air to enter the combustion chamber. As a consequence, the fires tend to burn much hotter. Also, and perhaps more importantly, much of the heat generated in the combustion chamber is lost up the flue pipe. Therefore, combustion stoves achieve efficiencies in the range of 50 to 60 percent, depending on the amount of time the stove is burned with the door open.

In an energy-short world, combustion stoves should be avoided. If you want to see the fire, purchase a radiant or circulating wood stove with a glass door, a feature offered by most wood stove manufacturers.

SHOPPING FOR AN EFFICIENT, CLEAN-BURNING WOOD STOVE

If you live in the country and have a ready supply of wood, or if you live in an urban or suburban setting and have access to lots of wood, a wood stove may be a very wise investment. As is the case with all renewable technologies discussed in this book, you will need to shop carefully. There are a lot of models on the market. So what do you look for?

I strongly recommend that you shop for efficiency. The higher the efficiency, the

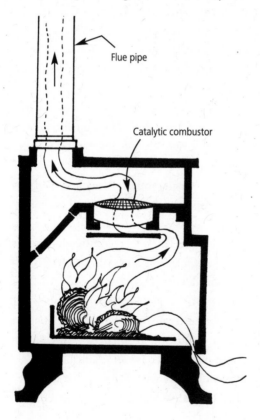

Fig. 6-5: *Catalytic burners in wood stoves increase their efficiency and reduce air pollution.*

Flue pipe

Catalytic combustor

better. As you might suspect, efficiency and cleanliness go hand in hand when it comes to a wood stove. A stove that wrestles as many BTUs out of the wood as possible also typically produces the least amount of pollution.

Be patient and research your options carefully. Visit as many dealers as you can and ask lots of questions. Be wary of claims that seem too good to be true.

Wood Stoves with Catalytic Converters

One of the first questions you will face when shopping for a wood stove is whether or not you should purchase a stove with a catalytic converter. To meet the clean air requirements of many cities, wood stove and fireplace insert manufacturers have equipped their products with catalytic burners (Figure 6-5). Catalytic burners contain a ceramic honeycomb structure coated in palladium or platinum (the catalyst). Like the catalytic converter in a car, the catalytic burner completes the combustion of unburned hydrocarbons — the gases escaping from the wood. (Remember, these gases ignite at a higher temperature than most wood stove fires provide, so unless there's a catalytic burner or combuster, the energy in these gases will be lost.)

Wrestling more heat from the wood by burning these gases boosts a stove's efficiency. Catalytic converters improve wood stove efficiency by 10 to 25 percent. Increasing efficiency not only means you get more heat from the wood you buy or collect, it means you will burn less wood and spend less time and money heating your home. It also reduces creosote buildup and the problems it creates. According to several sources, catalytic burners pay for themselves in two years, give or take a little.

Catalytic burners are located in chambers above, behind, or below the main combustion chamber. Because catalytic burners burn the gases and liquids that escape from firewood, these devices reduce creosote deposits on flue pipes by 80 percent or more, reducing the risk of house fires. Less creosote buildup also reduces cleaning costs.

Although catalysts are a great idea, they do have some drawbacks. One of them is that they don't last long. The life expectancy of new catalytic burners is only about three to six years. Yet another disadvantage is that they don't work well at higher elevations (according to the wood stove dealer near my former home in the foothills of the Rockies in Colorado).

Catalyst-Free Stoves

Fortunately, many wood stove manufacturers have found an alternative to the catalytic converter, one that's cheaper in the short term and the long term. These designs

How Wood Stove Catalytic Converters Work

Combustion is a chemical reaction between oxygen and organic materials triggered by heat.

Combustion temperatures inside a wood stove are normally only around 400° to 900°F (204° to 482°C), well below the normal combustion temperature of hydrocarbons. Wood stove catalytic converters dramatically lower the ignition temperature of hydrocarbons released from the fire from 1100°–1300°F (593°–704°C) to around 600°F (315°C). (Catalysts speed up the rate of chemical reactions by lowering the temperature at which they occur.)

Although catalytic burners start out burning at low temperatures, internal temperatures within a catalytic converter climb to 1700°F (926°C) or more pretty quickly. This is much higher than the temperature required to burn the gases and liquids released by firewood. Once the catalyst reaches these temperatures, it generates its own heat and will maintain temperatures sufficient to burn off hydrocarbons, even when the air supply to the stove air is restricted to prolong the burn time. Thus, even though a reduction in air flow creates a less efficient, cooler fire that releases more unburned hydrocarbons than a high-temperature fire, you don't need to worry about losing energy from your wood or polluting the atmosphere. Once the catalyst reaches temperature, it remains very hot.

utilize a baffle that forces combustion gases released from the fire — including unburned hydrocarbons — back over the

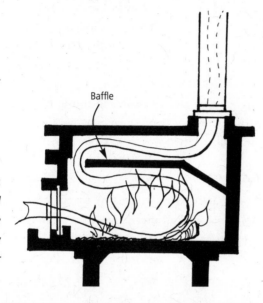

Baffle

Fig. 6-6:
Baffles in wood stoves increase their efficiency and reduce air pollution.

flames, causing them to ignite and burn (Figure 6-6). This process is typically referred to as *secondary burning*. Baffles increase combustion efficiency, in part, by increasing air turbulence. Greater air turbulence puts unburned hydrocarbon gases in contact with air and heat. The greater the mixing of these "elements," the greater the efficiency. Baffled wood stoves burn cleanly and achieve high efficiencies — on par with wood stoves equipped with catalytic combusters — and they're usually a bit cheaper.

What Else to Look for when Shopping for a Wood Stove

When shopping for a high-quality, clean-burning, energy-efficient wood stove, there are a number of features to look for besides those just mentioned — features that help you obtain the most efficient, safest wood stove on the market.

First on the list is durability. Durability is a function of the material from which a stove is made. Cast iron and welded steel are best. Stay away from stoves made from sheet metal. Although they are inexpensive, stoves built entirely out of sheet metal are typically designed only for occasional use and won't last very long if burned frequently. Thus, a sheet metal stove is a very bad investment for those who want to heat their homes with wood. Stoves made from one quarter inch thick, or thicker,

Should You Buy a Used Wood Stove?

Those inclined to save a little money may think about buying a used wood stove. Is this a good idea?

Not usually.

Although a used wood stove may be inexpensive, older models are often pretty inefficient. Being inefficient, they also produce a lot more pollution than newer models. Older models, for instance, produce 30 to 80 grams of particulate matter per hour. Newer models produce 90 percent less, or about 3 to 6 grams per hour. Even though you may save upfront, you'll burn more wood, spend more time tending the fire, and pollute more. You'll also need to clean your flue pipe more often.

steel plates that are welded together last a lifetime. Although the steel may warp a bit over time, steel stoves are durable and function year in and year out, without a problem.

Cast iron stoves, as just noted, are also a great choice. Some people consider them the best wood stoves on the market. Cast iron stoves do not warp like steel stoves and may last longer. However, despite its greater fire tolerance, cast iron is more brittle than steel. What this means to you is that a cast iron stove may crack if it is not shipped and installed with care. (Inspect a cast iron stove carefully for damage before firing it up the first time.)

Cast iron stoves can be coated with a durable, lustrous porcelain finish, adding beauty. However, cast iron stoves are typically the most expensive of all your options.

When shopping for a wood stove for your home, be sure to look for models that come with firebrick or ceramic tile linings. These materials shield the metal of wood stoves from high temperatures, increasing the stove's life span. They may also increase the temperature of the fire, which can increase combustion efficiency. You should also consider buying a wood stove that preheats the air delivered to the combustion chamber. Preheating usually occurs in pipes or channels inside the stove. They transport incoming air in

such a way that it can be warmed *before* it reaches the fire. This helps to maintain a hotter fire, and results in higher combustion efficiency.

Another highly desirable feature is a tight door seal. Good door seals help to prevent pollutants such as carbon monoxide from escaping from the combustion chamber and poisoning you and your family. A leaky door lets too much air into the combustion chamber, making it harder to regulate the combustion temperature and burn time.

How Big Should Your Stove Be?

When shopping for a stove, you will also need to consider its size. This depends on the size of your home and its energy efficiency, notably the level of insulation and air leakage. (Be sure to make your home as energy efficient as possible upfront.)

Fortunately, wood stove manufacturers offer a fair amount of information on their products that can help you make an intelligent choice. Unfortunately, there's no set standard, so manufacturers use different methods to rate their stoves. This makes it more difficult to compare products. Some wood stove manufacturers, for example, provide data on the heat output of their stoves, measured in BTUs per hour, while others list the number of rooms or the number of square feet their stoves are designed to heat. Still others list the cubic feet of room space their stoves will heat. Unless you know the precise conditions under which the stoves were tested (for example, the type of wood and the insulation levels in the test facility), data such as these may not be relevant to your home.

Personally, I think that your best bet is to ask a reputable local supplier who is familiar with the heating requirements in your area. Be sure he or she visits your home. If you have upgraded the insulation in your walls and ceilings, sealed your house, and taken other measures to reduce heat demand, be certain the supplier knows this. If not, you may end up buying a model that is larger than you need.

PROS AND CONS OF WOOD STOVES

Wood stoves are widely available and are easy to use. They are also relatively easy to install, although professional installation

Installing a Stove for Optimal Performance and Safety

Wood stoves need to be strategically placed for optimal performance and safe operation. Here are some general guidelines for installing a stove.

Choose a Central Location

When selecting a location for a wood stove, it's best to avoid placement against an exterior wall. That's because heat from the stove will warm up the interior surface of the wall, causing heat to flow out of the house. (Remember: heat flows from hot to cold.) Warming the inside wall increases the temperature difference across the exterior wall; the greater the difference, the greater the heat loss.

The best spot for a wood stove is in the center of the space you're trying to heat. A central location ensures that heat from the stove spreads naturally throughout the house by convection and radiation. If you are concerned that heat won't flow well, you can always install a ceiling fan or two to move heat around, provided that your ceilings aren't too high (over 16 feet). Be sure to locate a wood stove so that it can be easily and economically vented to the outside.

Protecting Against House Fires

Safe installation also requires placement of the stove and flue pipe the proper distance from combustible materials in floors, walls, and ceilings to prevent fires. Safe installation also requires the use of heat-resistant materials that shield the floor and the walls from the intense heat of the stove. Most stoves are installed on a tile base either purchased with the stove or made yourself. Tile, metal, or other noncombustible materials are often placed on the wall behind the stove. Details of safe installation are beyond the scope of this book, so be sure to check your local building code and recommendations from your insurance company. Professional installers know the codes and can save you the trouble of having to learn them.

by a competent crew is highly recommended (see the accompanying sidebar, "Installing a Wood Stove"). Many people choose wood stoves because they are one of the least expensive heating systems. You won't need the extensive and costly heat distribution systems required for conventional heating systems like forced hot air or radiant floor systems. Because there are so many suppliers and retail outlets in most locations, careful shoppers can usually get a pretty good deal on a wood stove, especially if they buy during the much slower summer season. Wood stoves are available in many different models, with many shapes, styles, and sizes to choose from. They can be used to heat a room or an entire home. The larger the home, of course, the larger the wood stove you'll need.

Further adding to their appeal, wood stoves require very little cleaning or maintenance. For an older model, a good, hot, cardboard fire once every few months or an annual cleaning of the stove pipe usually suffices to remove all creosote, greatly reducing the chance of a chimney fire that could spread to the home. New models require less cleaning because they burn so much cleaner, but don't get lazy. Flue pipes still need to be cleaned from time to time.

Many modern wood stoves are nice looking. Some models with glass doors allow you to view the fire, further adding to their appeal.

If you are concerned about air quality, remember that new wood stoves burn pretty clean. In fact, all new wood-burning stoves, except cook stoves, currently sold in the United States and Canada must comply with government regulations aimed at reducing urban air pollution. In the United States, the EPA sets standards for wood stoves. In Canada, the Canadian Standards Association has developed emission standards for wood stoves, fireplace inserts, and small fireplaces. Stoves certified by these agencies reduce smoke pollution by as much as 90 percent when compared to older models.

And, if you live in an area that imposes wood-burning bans on high-pollution days, you may be surprised to learn that EPA-approved stoves burn so cleanly that they are exempt from these bans. EPA Phase II stoves produce almost no smoke at all.

But what about carbon dioxide emissions from wood stoves and global warming?

Burning wood in a wood stove produces carbon dioxide to be sure. It's unavoidable. All organic matter that burns produces this greenhouse gas. However, unlike coal or natural gas or oil, wood is a renewable resource. As long as we plant trees to replace those that we cut down and burn, wood burning should not increase global carbon dioxide levels. (However, cutting wood with a chain saw and transporting wood in a vehicle powered by gasoline or diesel *will* release carbon dioxide into the atmosphere.) If you are cutting your own wood, replant trees. If you are buying from a local supplier, be sure to purchase from those who are good stewards of their forests and are actively involved in replanting, too.

Outside Combustion Air

Some states, like the state of Washington, require that all new wood stoves draw their combustion air from outside the home. This is designed to prevent backdraft — air being drawn down the chimney into the wood stove and then into the house when vent fans are being used inside a house. Why does this occur?

Vent fans create negative pressure inside the house that draws air in from any source it can. The flue pipe of a wood stove is a good source of replacement air. Backdraft of this nature may cause smoke and other pollutants from a smoldering fire to enter the home, creating a potentially dangerous situation.

To prevent this problem, combustion air for a wood stove can be brought in from the outside by creating an opening in the building envelope. It delivers cool outside air to the wood stove via a pipe. To be effective and safe, a system for supplying outside air must be carefully designed. Powered make-up air systems are probably your best bet. They are air systems that rely on a small fan to blow outdoor air into the stove. They only operate when the stove is in operation, and thus help prevent cold air from entering your home. Be sure to get professional help on this aspect of installation.

Another benefit of wood stoves is that they're fueled by a renewable resource. If sustainably managed, wood lots can provide a lifetime of fuel to a family without damaging the environment. (To learn ways to harvest safely and sustainably, you may want to read *The Good Woodcutter's Guide* by Dave Johnson.)

Harvesting woodlots can even benefit forests. For example, thinning woodlots reduces crowding and competition among trees for limited water supplies. Trees that are less crowded are healthier and better able to resist insects and other pests.

Wood burning is also economical. A cord of wood yields the same amount of useable heat as 200 gallons of heating oil, a ton of hard coal, or about 4,000 kilowatts of electricity. A cord of wood costs about $150, depending on your location and the type of wood. With home heating oil running around $2.80–$3.00 a gallon at this writing, 200 gallons of heating oil would cost about $560–$600. At 8 to 10 cents per kilowatt-hour, 4,000 kilowatts of electricity would cost $320–$400.

Wood burning does have some downsides. First and foremost, heating with wood may require a lot of work, especially if you are cutting from your own woodlot. You'll have to fell dead trees, trim off limbs, cut up the wood, haul it to the house, split it, and stack it. Even if you purchase unsplit logs that are delivered to your home, you'll be spending a lot of time cutting and splitting firewood. And don't forget, you'll be hauling wood into the house on cold winter nights, lighting fires, and tending to them. You'll be cleaning up bark and debris that inevitably drop on your floor and cleaning the ashes out of the wood stove every week or so, and then disposing of them. (Cooled ashes can be spread on your property to return inorganic nutrients to the soil.)

Although hard work is good for the body, and many people enjoy it, especially if it helps increase self-reliance, wood heat is more than many people want to deal with. You may want to consider purchasing firewood. Many companies deliver it to your home and some will even stack it for you — for a price, of course. Or you may want to consider a pellet stove, discussed shortly.

But those are not the only downsides of wood burning you should be aware of before you set off on this venture. As environmentally benign as wood is, it can cause indoor air pollution. As noted earlier, smoke may escape from a wood stove as a result of backdraft. Smoke may also escape when a stove is improperly opened, or from leaks, and can result in unhealthy indoor air. In recent field tests of Canadian homes, varying degrees of combustion spillage from assorted furnaces, fireplaces, and wood stoves were detected in an alarming percentage of the homes tested. Soot deposits on walls, ceilings, and drapes are not only a nuisance, they are a sign that the air is polluted with potentially harmful particulates. Wood stoves, even clean-burning models, contribute to ambient air pollution.

Yet another potential problem is that wood stoves tend to produce really hot, dry heat that may lead to uncomfortable interiors. This can be rough on sinuses and nasal passages.

Wood stoves also tend to create a heat gradient in a home. The hottest room is the one in which the stove is located. Outlying rooms are typically much cooler. Because most wood stoves don't come with fans to force air out and away from the stove, you may need to provide some way to better distribute heat. Many homeowners use ceiling fans or strategically placed small portable fans or box fans.

Wood heat may also be problematic in homes with cathedral or vaulted ceilings, a feature in many newer homes. Tall ceilings are a problem because heat from a wood stove rises and accumulates near ceilings — far away from the occupants. Although ceiling fans help to force the warm air back down, you will very likely end up burning more wood to achieve a comfortable room temperature. The more wood you burn, the more it costs, and the more pollution you produce. Second-story rooms or lofts may overheat, too, as a result of this phenomenon.

Remember, too, that because wood stoves require tending, they can't supply backup heat when you are away from home for any length of time. Only the automated pellet stoves will keep a home warm when you're away — so long as the pellet supply in the hopper lasts. (More on this shortly.)

In new construction, building departments rarely approve a wood stove as a primary heat source. You will most likely need to install an automatic wall heater, a furnace, or some other conventional heat source controlled by a thermostat. In existing homes, that's not a big deal. You probably already have a conventional heat source that you're trying to marginalize by using wood heat. Your furnace or boiler can be relegated to serve as a backup heat source.

Wood stoves can also be a fire hazard if improperly installed and maintained. Unfortunately, many houses go up in flames each year as a result of either poor installation or inadequate stove maintenance.

Another important consideration is that wood stoves are not as efficient as other forms of backup heat. For example, a wood stove isn't as efficient as a high-efficiency furnace or boiler. That said, wood is a renewable resource, while oil and natural gas, two conventional home heating fuels,

are not. Even though wood stoves may not be as efficient as conventional heat sources, they are still cheaper. To learn more, you may want to read John Gulland's piece, "Responsible Wood Heating" in *Home Power* magazine, Issue 99.

WOOD FURNACES

Another renewable option for home heating is a wood furnace. Wood furnaces are typically installed in basements, garages, or even, as you will soon see, outdoors. Wood furnaces can be used in conjunction with a conventional home heating system, for example, radiant floor heat, baseboard, or forced-air heating systems. The Lynndale and Yukon Eagle wood furnaces, for instance, contain blower fans that move air generated by burning wood in them through the ductwork of forced-air heating systems in homes (Figure 6-7). Some companies also design their furnaces to preheat domestic hot water. During the winter, these units not only heat the home, they provide hot water for washing dishes and other household tasks, greatly reducing natural gas or electricity consumption.

When shopping for a wood-burning furnace, be on the lookout for various features that improve combustion efficiency. Combustion air blowers, for instance, are fans that force a stream of air into the fire, increasing combustion efficiency. Secondary combustion chambers, like those found in wood stoves, also increase efficiency. A secondary combustion chamber increases efficiency by burning gases released from wood during combustion. In some furnaces, secondary combustion chambers are designed so that they have their own air supply, that is, a supply of air in addition to that provided to the main combustion chamber. This feature also promotes greater combustion efficiency.

Increased efficiency, of course, means you'll need less wood to heat your home. It also means your furnace will produce less pollution. Cutting down on pollution (soot and unburned hydrocarbons) helps keep the air in our cities and towns cleaner, and also has a very direct benefit to homeowners: it reduces creosote buildup in the chimney, reducing fire potential. This, in turn, reduces the need to clean the chimney as often, saving you money in the long run.

Some wood furnaces are designed to be used in conjunction with conventional fuels, ensuring that your home will be heated when you are gone for extended periods of time. Summeraire, a company in Peterborough, Ontario, Canada, for instance, manufactures a wood furnace that can be used in conjunction with home heating oil, natural gas, or propane. This dual-fuel furnace helps to prevent pipes from freezing when a homeowner is gone for extended periods.

YUKON EAGLE FURNACES, ALPHA AMERICAN CO.

Wood furnaces offer many advantages. They are pretty efficient and clean-burning. Some, like the wood furnace in the residence at The Evergreen Institute, take logs up to five feet long, which reduces labor. Wood furnaces also use a renewable resource and are fairly easy to operate.

On the downside, wood furnaces are much more costly than wood stoves and more difficult to install. They also require large fans to distribute hot air through the duct work of a home. Fans consume electricity, and they are often pretty noisy. And don't forget, you will need to periodically load wood and remove ashes from the furnace. Wood furnaces, like many home heating technologies, also present some

Fig. 6-7:
The Lynndale wood furnace.

fire hazard if not properly installed and maintained.

Those who like the idea of a wood furnace but don't have room to install one in their home — or don't like the idea of hauling wood to the basement during the winter — may want to consider an outdoor wood furnace.

A handful of companies produce outdoor wood-burning furnaces. Central Boiler in Greenbush, Minnesota, for instance, manufactures a line of high-efficiency outdoor furnaces. They're made from heavy-gauge carbon steel and titanium-enhanced steel (Figure 6-8). These

Fig. 6-8: *Central Boiler's outdoor wood furnace supplies hot water to the house that can be used in conjunction with radiant floor, baseboard hot water, and even forced-air heating systems.*

CENTRAL BOILER

furnaces look like a shed and can be placed as far as 500 feet away from a home or business. They burn wood to heat water, which is then pumped to the building through buried insulated pipe.

Outdoor furnaces are typically used in conjunction with radiant floor, baseboard hot water, or forced-air systems. However, they can also be used to supply domestic hot water in the winter, when they're running day in and day out.

Another well-made outdoor wood furnace is manufactured by HEATMOR in Warroad, Minnesota. These furnaces come in a variety of colors to match your house or to blend with the environment. Their furnaces are easy to load and maintain, and efficient to operate. They use a heavy-gauge steel that the manufacturer claims long outlasts its competitors'.

PELLET STOVES

If cutting firewood and hauling wood into your house all winter long isn't your cup of tea, you may want to consider a pellet stove (Figure 6-9). Pellet stoves are much like wood stoves, but easier and cleaner to operate. (I think of pellet stoves as the lazy man's — or lazy woman's — wood stove.) Pellet stoves are free-standing stoves that provide space heat. They burn wood, but not bulky pieces of firewood. Instead, they are fueled by dry, compressed wood pellets. The wood pellets are made from sawdust that is a waste product of the timber industry.

Sawdust, from milling trees to produce finished lumber, was once burned at local wood mills throughout North America in huge incinerators. They smoldered for days, producing tons of smoke that dirtied the skies of many otherwise pristine rural areas. Today, many of these facilities convert their waste into small pellets, resembling rabbit feed, and sell it to eager customers throughout Canada and the United States. This trade has not only helped bolster local economies, it has provided a clean-burning fuel source for millions of people across the continent.

Pellets are packaged in 40-pound bags and are sold in hardware stores and many large discount stores. You can purchase a single bag or buy pellets by the ton to save money. At home, pellets are typically stored in a dry place, either in the garage or basement or under a waterproof tarp in the backyard. Pellets are loaded into a hopper that automatically feeds them into the combustion chamber by a screw auger powered by electricity.

Because pellets are dry, and because they are fed into the combustion chamber at a controlled rate with plenty of air, these stoves burn very efficiently and cleanly. In fact, they're cleaner than most wood stoves.

Pellet stoves offer many of the advantages of a wood stove with fewer downsides.

SHANE KELLY AND DELL-POINT

As just noted, they burn a renewable fuel, which is a waste product. By burning wood pellets, you're helping reduce pollution in rural areas where mills are located. You're also helping to support the wood products industry.

Another advantage of pellet stoves is that the operator can set the rate at which pellets are fed into the combustion chamber. This regulates the heat output of the stove. Yet another advantage of pellet stoves is that they come with fans that blow hot air from their surface into the room.

Fig. 6-9: *Pellet stoves are the lazy person's wood stove. They're clean burning and operate automatically.*

This helps distribute heat throughout the house.

Pellet stoves are easier to load and not as messy as wood stoves. Another huge advantage is that they operate automatically, providing heat while you are gone, although usually no longer than a day or two. Many pellet stoves can be thermostatically controlled. And pellet stoves also allow for more precise control over heat production than wood stoves.

On the downside, pellet stoves require electrical energy to operate the auger and the blower fans that come standard on many models. In addition, pellet stoves usually cost more than similarly sized wood stoves, and they require more service. Wood pellets cost more than firewood,

too. You'll generate a lot of plastic waste from the empty bags, but you can use these for trash bags, if you like.

MASONRY HEATERS

Between about 1550 and 1850, Europe was gripped by an intense cold spell. Historians, in fact, often refer to this period as the *Little Ice Age*. At the beginning of the Little Ice Age, homes and buildings in Europe were primarily heated by wood. However, wood was burned in inefficient fireplaces, simple stoves, and braziers. Inefficient wood-burning, in turn, led to widespread shortages of wood. But like many problems, this one led to a happy ending. Shortages led to the development of masonry heaters, highly efficient wood-burning stoves that greatly cut back on wood use. Not only did masonry heaters use less wood, they also produced greater comfort. Masonry heaters were so successful that, by the end of the Little Ice Age, nearly every building in Europe was heated by one (Figure 6-10).

Masonry stoves are still in use in northern Europe. Not just in old buildings, either; they're often installed in new ones. In Finland, for example, 90 percent of all new homes are heated with masonry stoves — thanks, in part, to generous tax incentives offered by the forward-thinking and conservation-minded government. Masonry heaters are also very popular in

Fig. 6-10: Masonry heaters burn wood efficiently and provide a steady supply of heat. The masonry heater is an ideal heating system for homes, though they are difficult to install in existing homes.

ALBERT BARDEN

Norway, Sweden, Denmark, and Germany, where they are highly regarded for their excellent efficiency, reduced pollution output, and safety. Because they offer so many benefits, masonry heaters are even starting to make their way into new homes in North America as primary and secondary heat sources. I think they could be an essential element in our transition to a renewable energy future, especially in new housing stock in rural areas, for reasons you'll soon understand.

What is a Masonry Heater?

Masonry heaters are massive wood stoves typically made from bricks and mortar (Figure 6-11). Many masonry stoves are free-standing structures that are located in the center of the heated space so they can radiate heat outward in all directions. In some homes, masonry heaters serve as room dividers, providing heat to rooms on either side of them. Although many masonry stoves are free-standing (the most efficient installation), some stoves are built against outside walls.

Masonry heaters are designed to burn much hotter than wood stoves, which enables them to wring more heat from wood than even the most efficient wood stoves. They therefore require much less wood than a standard wood stove. Masonry heaters are also designed to emit a lower temperature heat, and release it steadily

over many hours. To understand how all of this is possible, let's peer inside a masonry heater to see how it works.

The main reason masonry heaters operate so efficiently is that their fireboxes are designed to burn wood at extremely high temperatures by ensuring ample air flow. Most designs also maximize turbulence in the firebox, which also increases combustion efficiency. As in wood stoves, the combustion chambers are lined by firebrick. Firebrick protects the rest of the masonry heater from the intense interior temperatures and also provides a greater level of insulation. This, in turn, holds heat in the firebox. By preventing heat from

Fig. 6-11:
The super-efficient masonry heater relies on three features: a well-insulated combustion chamber, a labyrinth flue, and high mass.

dissipating from the combustion chamber, internal temperatures can get quite high — in the range of 1200° to 2000°F (650° to 1095°C).

High-temperature combustion in a masonry heater results in the complete, or nearly complete, combustion of wood, including all of the gases and liquids driven out of a piece of wood after it catches fire. Masonry heaters, in fact, achieve efficiencies ranging from 88 percent to nearly 95 percent. (Wood stoves achieve efficiencies in the range of 70 to 80 percent.)

A hot-burning fire results in greater efficiency and also eliminates creosote buildup in the flue system. And it reduces air pollution. Like Phase II wood stoves, masonry stoves burn so cleanly, they're approved for use when wood-burning bans are in effect, even in several areas in the United States known for tough wood-stove standards, like Colorado, Washington, and San Luis Obispo County, California.

Another important feature of masonry heaters is the often huge amounts of thermal mass provided by the masonry materials from which they're built. These stoves are the leviathans of the heating world, weighing from one to eight tons.

What good is all of this mass?

Mass absorbs heat produced in the combustion chamber and slowly radiates it into the neighboring rooms. In fact, a masonry stove will release heat from a single load of wood for 6 to 24 hours, depending on the mass of the stove. Let me repeat: a single load of wood (40 pounds, or two armfuls) will heat a room or a small, open house for 6 to 24 hours!

Mass is not all that's required to make this happen. Masonry heaters are also designed so that hot combustion gases escape through a labyrinth flue system within the mass, as shown in Figure 6-11. In this design, hot gases from the firebox are directed downward through the mass into channels usually on both sides of the combustion chamber. As the gases flow through these channels, heat is absorbed by the masonry material. The heat slowly migrates through this mass, eventually reaching the surface of the stove. In a wood stove, hot gases escape straight up the flue pipe.

Unlike a wood stove, the surface temperatures of which reach 500°–700°F (260°–370°C), the surface temperature of a masonry heater is 155°–175°F (68°–80°C).

How Clean Are They?

Masonry stoves are in the super-low emissions category of wood-burning devices because they release only 1 to 2 grams of particulates per hour. Many wood-burning stoves release 3 to 6 grams per hour, although some of the best models achieve emissions of 1.5 to 3 grams per hour.

The brick radiates heat slowly into the room, creating long-lasting comfort with very little fuel. As the folks at Temp-Cast, a leading manufacturer of masonry heaters note, the masonry heater is "cherished for its gentle heating nature."

Masonry heaters wring lots of heat from wood by virtue of their super-efficient combustion chambers. They trap that heat in their mass, then slowly release it into the room, providing long-lasting heat. Another secret of the exceptional performance of a masonry heater lies in the way it heats rooms through its release of radiant energy. That is to say, the heat given off by the stove radiates from its surface outward, like heat from the glowing embers of a fire, warming the walls, floors, furniture, pets, and people. All solid objects and living beings in the vicinity are heated.

Should You Consider a Masonry Heater?

Masonry heaters are ideal for new home construction, but not so good for retrofitting. As noted earlier, masonry heaters weigh considerably more than wood stoves — an awful lot more! Bill Eckert, who once sold masonry heaters in Colorado, noted that his Temp-Cast stove kits weighed 2,800 pounds. When facing was added, the finished stoves ranged from 4,000 to 8,000 pounds — 2 to 4 tons. So, in order to retrofit a home, you would very likely need to fortify the floor to support the mass of the stove. For models placed on wood floors, for instance, you would need to install additional floor joists — maybe even steel beams — and very likely even additional support posts in underlying rooms. Of course, beefing up the framing will require the expertise of a structural engineer and the efforts of a skilled framer. Both add to the cost of a retrofit.

If you are installing a masonry heater on a concrete floor, you will very likely have to dig it up and pour a deeper slab to support the weight of the heater. Most concrete floors aren't strong enough to support a masonry heater.

For these reasons, very few masonry heaters are installed in existing homes. Most are installed in new homes designed specifically to accommodate the mass.

What's in a Name?

Masonry stoves or masonry heaters go by several different names, reflecting either the origin of their design or the type of stove. Those of Russian origin are often called Russian fireplaces or Russian stoves. Swedish masonry stoves are … you guessed it … called Swedish stoves. Those originally made in Finland are Finnish stoves. Some lighter-weight units are known as tile stoves because of the use of decorative tile as a facing material. Austrian, German, and Swiss tile stoves can be quite ornate.

— Dan Chiras, *The Solar House*

How Big a Stove Do You Need?

As a general rule, the colder the climate, the greater the stove's mass needs to be. The reason for this is that the more thermal mass a stove has, the longer it will heat. In extremely cold climates, five-ton stoves are common. In warmer climates, lower-mass models work well.

Those interested in masonry heaters can learn more about them and view some of their options by visiting the Masonry Heater Association's web page, mha-net.org. Be sure to check out their photo gallery; it showcases a variety of beautiful masonry heaters from different installers. While you are visiting this site, check out the various dealers' websites for photos of custom-made masonry stoves and kits.

Operating a Masonry Heater

Most masonry stoves are fired once or twice a day, using 35 to 50 pounds of wood, depending on the stove design, for each firing. Because there's a lag time between the time you start a fire in the stove and the time the heat reaches the surface of the stove, masonry stoves are not considered a fast source of heat. Hence, operators must anticipate their heat demand and plan accordingly. Firing mid-to-late afternoon, for example, will begin to provide heat for a family within a few hours, depending on the mass of the stove, and will keep a house warm throughout the night and into the next day.

Can You Install a Masonry Heater Yourself?

Homeowners can purchase and install masonry heaters themselves, but it's not a route I recommend. Masonry stoves require considerable knowledge of masonry materials and considerable expertise. Most stoves are built by skilled masons with years of experience in masonry stove construction. (To learn more about masonry heaters, I highly recommend David Lyle's book, *The Book of Masonry Stoves*. Kate Mink has also published an excellent article on masonry heaters titled "Living with a Masonry Stove," in *Home Power*, Issue 103.)

Pros and Cons of Masonry Stoves

As you can probably tell, I'm a big fan of masonry heaters. That said, I do recognize that they have some negatives. Let's take a look at both the pros and cons so you enter this adventure with eyes wide open.

To begin, as noted above, masonry heaters are efficient and clean-burning. The heat is then radiated into the house over a long period, providing hours of comfort from a relatively small amount of wood. In addition, because masonry stoves burn cleanly, creosote buildup in their chimneys is not a problem, nor are creosote fires.

Masonry stoves are also pretty attractive. Facings of brick, stone, adobe, tile, or stucco give them a distinctive look. In fact, masonry heaters can be blended very nicely with almost any architectural style and decor. There aren't many heating systems that you'd want to prominently display in your living room. This is one of them.

Many masonry stoves can be constructed with built-in bread or pizza ovens, so you can bake in them at the same time you are heating your home. Glass doors can also be installed so you can watch the fire. Note also that some designers build stoves that also heat water for radiant floor heat and hot water for showers and other domestic uses.

Masonry stoves require very little maintenance. An annual chimney cleaning is advised, although this is very likely a waste of time, or so say the experts. Creosote should never be a problem — unless the stove was improperly designed or built, or the user isn't operating it correctly. The most common mistake occurs when users build small, smoldering fires instead of hot, intense ones. Smoldering fires produce a lot of smoke, as do fires that burn green wood. Inefficient combustion, in turn, leads to creosote buildup. If you want less heat, build a small, hot fire that burns fewer minutes.

Building Your Own Masonry Heater

You may be tempted to build your own masonry heater. Although this is possible, as noted in this chapter, it takes a lot of skill. Stoves need to be airtight, and the materials must be capable of withstanding extremely high combustion temperatures. Perhaps even more importantly, masonry heaters need to be built to accommodate the expansion and contraction of the mass as it heats and cools. An owner-builder is unlikely to have the kind of skill and expertise to pull this off. Only the most experienced stone masons should be hired, and even then only those with experience building masonry heaters. When facing material such as tile or brick is installed, it must be spaced properly so it doesn't crack as the stove expands and contracts. Temp-Cast masonry heaters incorporate a "floating firebox" in their products. It isolates the heater core from the external masonry facing, which prevents the expansion and contraction of the firebox from damaging the facing. If not built correctly, or not built from the right materials, severe cracking can occur, leading to the collapse of the heater.

As also noted earlier, masonry stoves require additional framing of wood floors and additional reinforcement of concrete slabs, which add to the cost of new construction and make it difficult to add a masonry heater to an existing home. In airtight homes, make-up air needs to be supplied to masonry stoves. This also makes it difficult to retrofit a home for a masonry stove. And it adds some expense when building a new home.

Small amounts of fly ash may need to be removed periodically from the masonry flues. Accordingly, a masonry stove should have a clean-out that permits access to the smoke channels.

On the downside, masonry heaters are not widely available. To find a dealer near you, contact the Masonry Heater Association of North America's web page listed above, or ask around for a dealer. A local wood stove retailer may be able to help you out. They may even sell masonry heater kits that they can install for you.

Like wood stoves, masonry heaters aren't generally controlled automatically. Thus, they don't provide heat when you are away from your home for any length of time, although some models can be equipped with a thermostatically controlled natural gas burner for such times.

Masonry stoves are not cheap, either. Building a masonry heater yourself could cost $5,000 or more. If you have a professional install one, expect to pay $6,000 to $15,000, give or take a little, depending on the size of the stove and the options you select, for example, a built-in pizza oven.

In closing, although it may seem out of step with the times to heat a house with a 16th-century wood-burning technology, don't forget that we descend from many generations of people who heated their homes with wood.

Wood is a renewable resource — one we can count on from year to year if we manage our forests sustainably. And, as I like to remind audiences, many old ways of doing things are still the best ways.

CHAPTER 7

PASSIVE COOLING

STAYING COOL NATURALLY

One hot July day, shortly after I'd moved into my new energy-efficient home in the foothills of the Rocky Mountains, a white car pulled into my driveway. Two women emerged, dressed in business suits and carrying clipboards. I guessed that they were representatives of some branch of government.

I greeted them at the door and learned that they were from the county tax assessor's office. They had arrived unannounced to look over my newly-built home to take measurements and to list its various features so they could determine my property tax. After stepping out of the 95-degree summer heat into my naturally cooled home, one of the women glanced at the checklist on her clipboard and said, "You must have central air." "No,"

I said, "this house is naturally cooled." She shot me a suspicious look. "Really," I said, "it's super-insulated and earth-sheltered. Together, they keep it pretty cool in the summer. There's no need for air conditioning." She gave me another look, this one a little more impatient. Her impatience, no doubt, stemmed from frequent conniving of members of the general public who want to dodge their duty to pay taxes. I don't particularly like taxes, but I realize that if we're going to have decent education, snow-free roads, police protection, fire protection, and other vital services, we've got to pay for them. Undaunted, I repeated my earlier statement, then added, "Look around, there's no air conditioner." With that, the ladies went about their business, measuring rooms and checking off various features of the home and — no doubt — secretly searching for ducts that deliver cool air to my home. Outside, they scoured the landscape — no doubt looking for my nonexistent central air conditioner. Cooling the various homes I've owned or rented over the years has been a challenge. But this house, designed to be passively cooled, doesn't have an air conditioner, nor do we need one, despite some pretty hot summer days.

You may face a summer cooling challenge, too, which is only bound to worsen as global warming intensifies. Fortunately, you don't have to build a new home to meet the challenge. With the time-tested techniques outlined in this chapter, you can retrofit your home to greatly reduce — perhaps even eliminate — your need for mechanical cooling. All you need to do is apply the principles and practices of the lost art of passive cooling.

WHAT IS PASSIVE COOLING?

Passive cooling is a key element of a larger strategy known as *natural conditioning* — heating, cooling, ventilating, and lighting a building naturally, that is, without mechanical or electronic devices and without outside energy.

Like passive solar heating, passive cooling may require some backup from time to time. The goal, however, is to reduce our reliance on mechanical cooling and

ventilation systems and the outside energy needed to run them. In the process, we slash our energy bills, increase our energy independence, and dramatically reduce our impact on the environment, the life-support system of the planet and, lest we forget, the source of all our wealth.

How Does it Work?

Passive cooling taps into natural forces, such as cool breezes, shade, and cool nighttime air as well as ordinary building components, such as insulation, overhangs, and energy-efficient windows. As noted in Chapter 4, many of the steps taken to heat a home passively also contribute to passive cooling. When building a new home, for instance, the simple act of orienting a building to true south increases wintertime passive solar gain while greatly reducing summertime heat gain. The net effect of this simple measure is that the house stays warm in the winter and cool in the summer, naturally. Other passive heating strategies like sealing leaks in the building envelope, insulating well, installing energy-efficient windows, and building with sufficient overhangs also enhance year-round comfort. But there's a lot more you can do to passively cool a new home, and there's much you can do to an existing home to reduce its reliance on mechanical cooling and the costly, environmentally damaging fuels that power it.

Moreover, it doesn't matter where you live. Passive cooling techniques work well in all climate zones, from hot, humid regions like the midwestern and southeastern United States to hot, arid climates like the western and southwestern United States, although humid climates pose greater challenges, as you shall soon see.

Why is Passive Cooling Important?

Most homes in the United States are cooled by air conditioners powered by electricity derived from the combustion of coal or nuclear fission. A fair amount of natural gas and a tiny amount of oil is also used each year to generate electricity. Passive cooling could help reduce our dependence on nuclear energy and fossil fuels and reduce the ever-growing strain on North America's already-taxed electrical supply system. This, in turn, will help to reduce the potential for annoying, costly, and sometimes highly disruptive summer brownouts and blackouts.

TOOLS IN THE PASSIVE COOLING TOOLBOX

Passive cooling relies on numerous seemingly insignificant measures that, when combined, dramatically reduce cooling loads in a building. These measures can be grouped in four general categories: (1) reducing internal heat gain, (2) reducing external heat gain, (3) purging built-up heat, and (4) cooling people directly. Let's take a look at each one.

REDUCING INTERNAL HEAT GAIN

All homes contain numerous internal sources of heat. These, in turn, result in a phenomenon called *internal heat gain*.

Common sources of internal heat gain include people, pets, electronic devices, and lights, which all generate varying amounts of heat. Even small transformers for answering machines produce a small amount of waste heat. The most significant sources of heat include conventional stoves and ovens, clothes dryers, dishwashers, water heaters, incandescent lights, aquarium lights and heaters, television sets, and computers. Even microwave ovens produce waste heat, though much less than conventional stoves and ovens.

When added up, these numerous small and seemingly insignificant internal heat sources collectively generate a substantial amount of heat. In the winter, internal heat sources help keep buildings warmer; in fact, in some super smart, energy-efficient office buildings, internal heat sources from copy machines, computers, lights, people, and other sources satisfy much of the wintertime heat load. At times, internal heat sources are an asset.

In the summer, however, internal heat sources become a liability; they make our homes and offices warmer and less

When You're Hot, You're Hot

An adult produces as much heat as a 50- to 70-watt lightbulb, depending on his or her age, metabolic activity, and level of physical activity.

comfortable, and they contribute to higher fuel bills. Although ridding a home of internal heat sources could increase heating costs, cooling costs often outweigh heating costs. It therefore makes economic sense in the long run to eliminate internal heat sources.

Eliminating internal heat gain is the first step to passively cooling your home. You can start by systematically locating all sources of internal heat gain in your home. To begin, take out a piece of paper and make a list of all appliances, electronic devices, and lights in your home. Which ones are used most frequently? Star those items and focus your attention on them.

When your list is completed, jot down a strategy or two for reducing or eliminating the heat generated by each item on your list. For example, next to lights you could write "replace with compact fluorescent lightbulbs." Next to the stove and oven, you can write "use the microwave during the summer more than the stove." Microwaves use much less energy to cook food and thus produce less waste heat. You could also consider cooking more meals

outdoors on a grill or perhaps on a solar oven. Or, you could eat meals that require less cook time, for example, salads with chicken or boiled eggs for protein. Next to the water heater, write "install an insulated water heater blanket." Next to the clothes dryer, you could write "hang clothes on a solar clothes dryer" (commonly referred to as a clothesline).

As you work through the list, you will find that most heat-reducing solutions are fairly easy and inexpensive. Some may require more costly fixes — like replacing energy-inefficient appliances.

Replacing an old, inefficient water heater with a tankless model, especially a model without a pilot light, will reduce internal heat gain and save loads of money over time (see Chapter 2). Such a costly solution often makes good sense, especially if the appliance needs replacement anyway.

Table 7-1 lists major heat producers in a typical North American home and ways to reduce their contribution to internal heat gain — both the inexpensive measures and the more costly ones.

REDUCING EXTERNAL HEAT GAIN

Strategies to reduce internal heat gain are, for the most part, relatively painless and economical. Unfortunately, the most significant source of heat in most homes and businesses during the summer is external

	Table 7-1 Reducing Internal Heat Gain		
Heat Source	**Contribution to Internal Heat Gain**	**Cheap Option**	**More Costly Option**
Incandescent lights	Major	Use lights more sparingly. Turn lights off when not in use. Unscrew light bulbs in areas where they are not needed.	Replace with compact fluorescents. Install occupancy sensors that switch lights off when rooms are unoccupied.
Water heater	Major	Turn temperature down to 120°F (49°C). Install insulation: water heater blanket. Insulate hot water pipes.	Replace old models with on-demand (tankless) water heaters.
Stove and oven	Major	Eat more cold meals during the summer. Cook outside as much as possible. Use the microwave more during the summer. Bake at night. Run the exhaust fan when cooking.	Replace old, worn out gas stoves (with pilot lights) with models that have electronic ignition switches.
Clothes washer	Minor	Use the cold or warm water settings. Wash clothes at night.	
Computer	Minor	Turn the computer off when not in use.	Replace old, outdated computers with energy-efficient models.
Clothes dryer	Major	Hang clothes on outside line. Dry larger loads. Close off utility room to rest of the house. Open window to utility room when clothes dryer is in use.	Replace with a more energy-efficient model. ☞

Heat Source	Contribution to Internal Heat Gain	Cheap Option	More Costly Option
Television	Major to minor depending on daily use and size and efficiency of TV.	Watch TV more sparingly. Unplug TV when not in use. Plug TV into power strip and turn off when not in use.	Purchase the most energy-efficient model possible, when buying a replacement TV set.
Furnace (pilot light)	Minor	Turn off pilot during the cooling season; reignite during the heating season.	When replacing furnace, purchase an efficient model that does not have a pilot light.
Pets	Minor	Let pets spend more time outside, but be sure to provide shade and water for them.	
People	Minor	Spend more time outdoors on porches or patios or on shaded lawns.	
Shower	Major	Turn water heater temperature down. Take shorter showers. Open window when showering. Run exhaust fan when showering. Replace showerhead with a more efficient model.	
Stereo	Minor	Turn stereo off when not in use. Unplug components of stereo system that are not frequently used, for example, tape players. Plug stereo into power strip and turn off when not in use.	
Dishwasher	Major	Hand-wash dishes. If it is not already turned off, switch off the drying option.	

heat gain — heat that enters buildings from the outside.

Heat enters our homes in a variety of ways. For instance, light striking the roof of a house is absorbed by the shingles and converted to heat (Figure 7-1). Much of this heat is conducted through the roof into the attic. It then passes through the insulation or, more likely, through the wood framing of the ceiling and radiates into our homes. (Attics are typically 30° to 40°F [17° to 22°C] hotter than ambient air in the summer.) In homes built without attics, heat passes directly from the roof through the ceiling cavity into the living space.

Sunlight striking windows and walls also heats up our homes, as shown in Figure 7-1. Warm air surrounding our homes also contributes to external heat gain. Warm air is generated when sunlight strikes driveways, streets, sidewalks, patios, and decks. Warm air heats up the walls and windows of a building, much the same way a pie is heated in an oven. External heat is then conducted through walls. Warm air can also enter buildings through open doors and windows and through the multitude of cracks in the building envelope, as explained in Chapter 2.

Stopping the flow of heat into a building from the outside requires a new set of tools. One of the most important is to cool the external environment by

Sunlight strikes pavement and ground

Sunlight striking roof, walls, and windows flows into a house by conduction and radiation

Warms walls and windows

Warms air

Warm air leaks into home through openings in building envelope

Heat conducted to interior

DAN CHIRAS

Fig. 7-1: *The arrows indicate all of the potential sources of external heat gain for a typical house.*

providing shade, for example, by planting shade trees.

Reducing External Heat Gain Through Natural Shade

Shade trees cool in two ways. They block sunlight directly, reducing heat absorption by roofs, walls, driveways, walkways, and street pavement (Figure 7-2). They also cool evaporatively. Just like perspiration that cools our bodies on a hot summer day, evaporation from the surface of leaves removes heat from the environment, creating a cooler atmosphere. This happens because evaporation requires energy to move water molecules from within a leaf into the air surrounding it. That energy is supplied by heat. (Scientists call this the *latent heat of vaporization*, just in case you wondered.)

Evaporation occurring in trees and vegetation is one reason why it is typically three or four degrees cooler in rural areas surrounding cities than in the cities themselves. (Cities typically have much less vegetation than the surrounding countryside.) The other reason is that cities contain lots of unshaded, heat-absorbing surfaces such as sidewalks, parking lots, streets, and massive concrete buildings. Carefully placed shade trees can block sunlight that would otherwise shine down on homes and businesses as well as streets, sidewalks, driveways, patios, and other sunlight-absorbing, heat-producing surfaces, even sheds and nearby vacant lots.

Don't plant tiny seedlings to save a few bucks. Young trees are less likely to survive, and they will take many years to reach full size. You may be in the retirement home — or in the nearby cemetery — before they begin producing significant shade. Bite the bullet and purchase large trees or, perhaps, fast-growing trees like cottonless cottonwoods. (Consult your local nursery for advice on the hardiest, fastest-growing shade trees.)

I recommend planting deciduous as well as non-deciduous trees (evergreens) on the north, west, and east sides of a home to provide shade. If you are going to retrofit your home for passive solar to provide space heat, keep the south side tree-free. If you want to plant trees on the south

Fig. 7-2:
Shade trees cool homes by blocking sunlight and through evaporative cooling.

DAN CHIRAS

side, be sure to minimize the number and then plant only high-crowned deciduous trees; they'll lose their leaves in the fall which will still allow some solar gain (Figure 7-3). Certain species, like hickory, grow long, tall trunks with few branches and leaves near ground level. Instead, their branches and leaves are concentrated near the top of the tree. Thus, they can provide shade in the summer, and won't block as much winter sunlight as many species of tree. (Consult with a local nursery on the best type of tree.) Whatever you do, don't plant evergreen trees on the south side of your home (Figure 7-4). Evergreens dramatically reduce, and even eliminate, the potential for passive solar gain.

If you are planning on adding a solar hot water system or solar electric system, be certain that the roof is unshaded year round. Solar hot water systems are not as sensitive to shading as solar electric systems. As you will learn in Chapter 8, even a small amount of shade on most solar arrays will dramatically reduce electrical production.

If the roof is shaded, you can still install a solar hot water system or a solar electric system, but you'll need to have access to a sunny space near your home, for example, the unshaded roof of your garage or an unshaded area of your lawn where you can mount solar electric modules on a pole, tracking device, or rack (Chapter 8).

Fig. 7-3: *These high-crowned trees allows ample low-angled winter sun to penetrate the south-facing windows of the home from which the photo was taken, yet offer a fair amount of cooling shade in the summer.*

Reducing External Heat Gain by Shade Structures

Shade can also be provided by structures such as an arbor. As shown in Figure 7-5, an arbor is a wooden frame for growing

DAN CHIRAS

CATHERINE WANEK

Fig. 7-4: Evergreens on the south side of homes block the low-angled winter sun and dramatically reduce, or even eliminate, potential passive solar gain.

Fig. 7-5: Vine-draped arbors such as this provide many benefits and are a great passive cooling measure.

grapes, ivy, or other forms of vegetation (even squash or cucumbers) in the spring and summer. The vegetation may provide edible fruit and flowers or habitat for birds and butterflies, while blocking sunlight. Shade provided by an arbor reduces heat gain through the windows and walls of a home. In addition, the vegetation adorning an arbor also cools the air around a home (via evaporation), reducing external heat gain. Arbors create lovely spots for you to relax on a hot summer day, too.

Vines can also be grown on trellises or directly on the exterior walls of a home. Ivy adds a touch of beauty to a home, and also shades the walls and provides evaporative cooling. However, ivy can damage walls, especially masonry walls.

Appropriate overhangs also reduce external heat gain (Figure 7-6). Unfortunately, many homes in North America were built without overhangs to save money — and many continue to be built without them to cut costs. And some designs, for example, traditional southwestern-style homes, do not include overhangs.

Although it is possible to retrofit a home that lacks sufficient overhangs, it is not always easy, nor is it typically inexpensive. It may also detract from a home's aesthetics. Don't despair, though; there are ways to compensate for the builder's cost-saving measure, for example, awnings and external shades, which are discussed in the next section.

Reducing External Heat Gain via Mechanical Shading Devices

One way to provide shade to walls is by installing awnings like those shown in Figure 7-7. Properly installed awnings can reduce heat gain through windows by as much as 65 percent on south-facing walls and 77 percent on east-facing walls. Select light-colored awnings to reduce heat buildup around the house.

Awnings are either fixed or retractable. I recommend retractable awnings whenever possible, especially on the south side of a home, to ensure maximum solar gain in the winter. You may even want to consider motor-driven awnings with wind sensors that retract in high winds to prevent damage. For more information on retractable awnings go to retractableawnings.com. This website contains 3-D rendering software that allows you to enter a picture of your home so that you can see what different retractable awnings will look like before you lay your money down.

One of the most convenient options for reducing external heat gain is the common, ordinary window shade. You may already have them. If so, be sure to use them to block unwanted sunlight during the cooling season.

Fig. 7-6: *Overhangs protect windows and walls from the high and intermediate summer sun.*

Window shades come in a variety of forms and are especially useful on west-facing windows during the summer as the sun makes its long, slow descent. Window shades are also useful on the hot days of

Fig. 7-7: *Awnings can dramatically reduce heat gain through windows in the summer.*

spring and fall if sun is beginning to provide a little unwanted heat.

For best year-round performance in colder regions, give strong consideration to insulated window shades, for example, cellular shades or shades made from quilted fabrics or from Warm Window material. (Warm Window fabric comes with two layers of polyester insulation to reduce heat flow and a layer of silver Mylar to reflect heat.) Insulated window shades help keep the sun out during the late summer and will help hold heat in a building in the winter.

Window shades can be installed inside or outside (Figure 7-8). In the vast majority of homes, they are installed inside for practical reasons: accessibility, ease of operation, and protection from the weather. But while most window shading devices are internal, external shades are preferable.

Why?

Unless shades are coated with a reflective material, internal shades still allow considerable amounts of sunlight inside buildings. That sunlight is converted into heat.

External shades, on the other hand, block — or nearly completely block — sunlight from entering a building, which greatly reduces external heat gain. But installing external shades doesn't mean you have to live in total darkness. Shades can be drawn only for short periods, when sunlight penetration is greatest, for example, in the late afternoon. Or, if you are at home all day, shades can be drawn as needed, starting on the east side. When the sun moves to the south, shades along the south side of the home can be drawn, while east-side shades are opened. Later in the day, shades can be drawn over west-facing windows. Some types of external window shades permit some light to enter your home; these will be described shortly.

In recent years, manufacturers have devised some rather innovative internal and external window shades. One is the solar screen. Solar screens consist of a fine mesh that blocks sunlight much like the solar screen used in greenhouses in the summer to prevent overheating. Solar

Fig. 7-8: *Window shades on the outside are more effective than internal shades when it comes to reducing external heat gain.*

DAN CHIRAS

screens are installed in frames that are usually mounted in window frames from the outside.

Another is the solar shade, a tinted acrylic shade typically installed inside. Although the solar shade blocks sunlight, it is transparent enough to ensure you have an outside view. Another ingenious invention is the cellular shade that is raised from the bottom up. Cellular shades can be used to block sunlight from penetrating the lower parts of windows in late spring and early fall.

Don't despair if your home has metal blinds. They're also useful for reducing heat gain — and heat loss in the winter.

For more on window shade options, you may want to check out the chapter on passive cooling in my books *The Solar House* and *Green Home Improvement*.

Reducing External Heat Gain by Sealing Cracks, Upgrading Insulation, and Replacing Windows

Sealing up the cracks in a building envelope will help keep a building cooler in the summer. It's one of the least expensive, easiest, most profitable, and most often overlooked steps you can take to make a home or business more energy efficient. As you learned in Chapters 2 and 4, air sealing with caulk and weather stripping not only makes a home or business cooler in the summer, it helps keep them warmer

in the winter. You may want to refer to my book, *Green Home Improvement*, for details on specific projects.

Insulation in ceilings, walls, and around foundations also thwarts external heat gain — quite significantly. Because I listed many options for upgrading insulation in Chapter 4 on passive solar heating, I'll simply refer you to that section now, if you haven't already read it.

Suffice it to say that any step taken to improve the heat resistance of the building envelope will help keep the home cooler in the summer. Because of this, insulation is one of our greatest allies in reducing external heat gain.

Yet another dual-function measure — that is, a measure that helps make a home warmer in the winter and cooler in the summer — is window replacement. Replacing old, leaky, and energy-inefficient windows with modern, airtight, energy-efficient models can make a world of difference. As noted in a publication on home energy efficiency published by the US DOE, about 40 percent of summertime external heat gain enters a home through its windows. If you want to save energy and achieve a greater level of energy independence, low-e windows are a must! Even if you can't afford to replace your windows, other upgrades like internal and external storm windows, discussed in Chapter 2, can go a long way toward eliminating external heat gain and making your summers much more comfortable.

Reducing External Gain by Repainting Your Home

Several studies over the past 20 years have shown that light-colored paints dramatically reduce heat gain in urban homes. Light-colored walls help reduce heat gain because they reflect more sunlight off a building than darker colored walls. Lighter siding also lasts longer, so it saves you money, too.

Combined with other measures, like planting shade trees, sealing leaks, and upgrading insulation, light-colored paints can dramatically reduce cooling load — the amount of energy required to cool a home. One study in a residential neighborhood in Phoenix, Arizona, for example, showed that three trees planted near light-colored houses could cut cooling costs by 18 percent per year. In Sacramento and Los Angeles, the savings were even greater: 34 and 44 percent, respectively.

Combined with other measures, light-colored walls can take a significant chunk out of your annual fuel bill. Certain types and colors of roof shingle can also reduce external heat gain. If your home is in need of re-roofing, consider lighter-colored shingles, metal roofing, or Spanish tile if they fit in with your home's design and color schemes. Lighter-colored shingles reflect more sunlight back into the atmosphere, reducing heat gain through ceilings. Metal roofing tends to be installed in ways that create a small airspace between the roofing material and the decking, which retards heat gain. Spanish tiles create even larger airspaces that reduce heat absorption.

Reducing Heat Gain by Installing Radiant Barriers

Radiant barriers are relatively new products that are becoming more and more popular each year in new construction as well as in homes being retrofitted for energy efficiency, especially in hot areas like the southeastern United States (Figure 7-9 a and b). Radiant barriers are durable aluminum foils (with a sturdy backing material of paper or plastic). They are stapled to roof rafters or attached to roof decking. Radiant barriers block heat from entering the attic. According to the Florida Solar Energy Center, a radiant barrier can cut cooling costs by 8 to 12 percent per year in hot climates like Florida's. This small but significant reduction, when combined with other measures described in this chapter, can nudge your home much closer to energy independence and cheap, low-impact comfort. We installed radiant barrier in my office at The Evergreen Institute and noticed an immediate and dramatic effect. You can purchase radiant barrier online.

Purging Heat

In my house in Colorado, I succeeded in eliminating enough external and internal heat gain that I am 100 percent passively cooled. Much to the dismay of the tax assessor's office, I have no cooling system, yet stay cool all summer long, even when outside temperatures are in the mid- to high 90s. In most existing homes, built with less exacting energy standards, it is nearly impossible to achieve this. No matter how hard you try to eliminate internal and external heat sources in your home, some heat will seep in. You can live with the mild discomfort or find ways to eliminate it. Getting rid of that heat is the third line of attack in achieving passive cooling. What you do to address this problem depends on the climate in which you live.

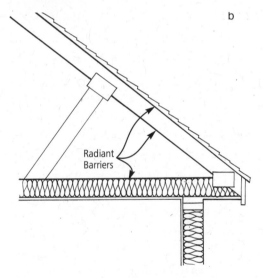

Fig. 7-9: *(a) Radiant barrier being installed by Rocky Huffman of Sustainable Energy Systems of Wichita, Kansas. Radiant barriers help reduce external heat gain through roofs and are becoming much more popular in modern construction. (b) This drawing shows three locations in which radiant barrier are installed.*

Purging Heat via Natural Ventilation

In temperate climates, heat can be purged naturally by establishing cross-ventilation — that is, by allowing air to flow through your home during the day. For this strategy to work, you must have a cool air reservoir. That is, you must have a cooler area, for example, a well-shaded backyard on the north side of the home, from which you draw cool air inside. Cool air entering through windows on the north side flows naturally through a house, purging heat. It then exits through windows on the same level on the warmer south, east, or west side.

Air can also be encouraged to flow vertically from windows on lower levels to windows in upper stories. Creating air flow from lower to upper levels occurs because of a phenomenon known as the *stack effect* or *chimney effect*. This results from a phenomenon we're all familiar with: hot air rises. When hot air in a home rises, it draws cooler air in through open windows on lower levels. If those windows open to much cooler outdoor air, you should be in for a treat. Natural convection currents should naturally purge heat from your home, helping to maintain internal comfort.

The natural flow of cool air through a home performs a second function as well: it helps to cool people down much like a fan blowing air across your face. (I'll discuss this shortly.)

Natural cross-ventilation and the stack effect produce a soothing flow of cool air that removes heat from your home. To enhance the effect, you may want to draw outside air into your home through your basement, provided the basement is not too musty smelling.

Basements tend to stay cool in the summer. If funneled through a cool basement, outside air can be cooled even more. Outside air, cooled by the basement, then flows up through the stairwell into the first level — and even into the second level of two-story homes — where it exits via a few open windows.

Natural ventilation works in virtually all climates early and late in the cooling season. There's generally no need to run air conditioners during this time, unless you need to remove humidity from the air. In some climates, though, natural ventilation can work throughout most, if not all, of the cooling season, especially if you have sealed the cracks, insulated well, planted trees for shade, painted your home a light color, and pursued other measures described in this chapter to reduce your cooling load. Hot, arid climates are one such place where passive cooling works well.

Nighttime Heat Purging

In desert climates, daytime cross-ventilation strategies work well early and late in the cooling season. For most of the rest of the spring, summer, and fall, however, daytime temperatures are too high for this strategy to work. Natural ventilation must occur during the night. Why?

In deserts, air cools dramatically at night because of the lack of moisture in the atmosphere. (Moisture in the atmosphere is a greenhouse gas that retains heat like a huge blanket over such areas.) Thus, even though temperatures may climb to 105°F (40°C) or more during the day, evening temperatures in the desert typically fall into the 50s and 60s (10° – 15°C).

The sudden and significant drop in evening temperature provides conditions optimal for nighttime heat purging. Opening a few windows to encourage cross ventilation, or to promote the stack effect in two-story homes, allows cool, relatively dry air to circulate through a home. The cool nighttime air scours heat that has accumulated inside the structure from internal and external sources. By morning the home is cool as a cucumber and ready to repeat the cycle once again.

Cooling buildings that contain significant amounts of thermal mass, such as tile floors or adobe walls, works even better in such climates. That's because these forms of thermal mass absorb heat during the day. This helps to maintain cooler interior temperatures. At night, cool outside air flowing through the house strips the heat from the thermal mass, rejuvenating it for another day's cooling cycle.

Not only are cool evenings in desert climates ideal for passive cooling, they're amazingly comfortable for sleeping. Unless you're in a highly developed area with so many homes that evening breezes are thwarted and so much heat-absorbing pavement that it takes a long time to cool down, nighttime purging, when combined with all of the other measures I've been discussing, could provide your ticket to energy independence.

A Little Active Boost

In some areas, you may have to rely on some active measures to provide a little boost to passive cooling to achieve comfort levels you and your family find desirable. Three of the least expensive and most effective measures are window fans, attic fans, and whole-house fans. We've had to employ these measures in my educational center in Missouri with its hot, humid summer days.

Box fans mounted in windows can dramatically increase air circulation; they're an inexpensive and simple way to enhance natural ventilation.

Attic fans are usually mounted on the roof, like the solar attic fan in Figure 7-10. Attic fans typically operate during the day to purge heat that accumulates in the attic. By doing so, you can reduce the amount of heat that enters through the ceiling.

Fig. 7-10:
Solar attic fans like this one being installed by students in my hands-on green building class at The Evergreen Institute purge heat from attics, reducing external heat gain through ceilings.

DAN CHIRAS

Whole-house fans mount in the ceilings, usually in a central location (Figure 7-11c). They draw air in through open windows, then vent it out through the attic. (Openings in the gable ends of homes allow air to escape from the attic.) I use our whole house fan in the faculty residence at The Evergreen Institute for a half hour or so at night to purge heat gained during the day.

"Whole-house fans are ideal for homes in climates that don't require air conditioning. However, they are also a perfect solution for homeowners who wish to run their air conditioning less often," says Marianne Methven, vice-president of marketing for Air Vent, Inc., a company that manufactures amazingly quiet whole-house fans. For instance, attentive homeowners use whole-house fans early and late in the cooling season — or early and late in the day — when daytime temperatures are below the temperature inside their homes; others use them at night to encourage nighttime purging. This strategy, however, won't work when outside temperatures remain high during the evening, as in hot, humid climates like those in Georgia and Louisiana.

Whole-house fans can be noisy. Responding to this problem, Air Vent, Inc. created Whisper Aire, the sleek, attractive model, shown in Figure 7-11c. According to Methven, homeowners said they liked the benefits of whole-house fans (energy efficiency, fresh air), but the fans were too difficult for the average do-it-yourselfer to install. To simplify the installation process, their unit is designed so that it can be installed between ceiling joists spaced at either 16 or 24 inches on-center, which means the fan will fit easily between joists. Because it's smaller and lighter than traditional whole-house fans, it's easier to handle during installation, too.

Another advantage of the Whisper Aire whole-house fan is its insulated automatic shutter. Providing an R-value of 25, the shutter minimizes the heat transfer between attic and living spaces. Whisper Aire is operated via remote control. This eliminates the need to drop wires down an interior wall to a switch, which makes this model even less complicated for the typical homeowner to install. The remote has five speeds and four timer settings that can be set to circulate for one half, two, four or six hours.

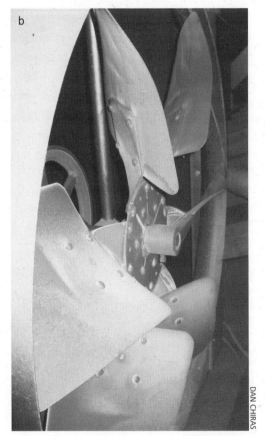

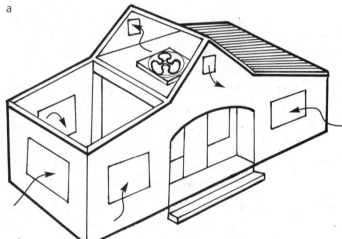

Fig. 7-11: (a) Whole-house fans create cross ventilation and also purge heat from attics. (b) In earlier years, large, noisy whole house fans like this one were installed. (c) Sleeker, more efficient, and much quieter models like this one are now available, and are relatively easy to install.

Air Conditioners and Evaporative Coolers

Heat can also be purged at night — or on days when passive cooling measures can't keep up with internal and external heat gains — by air conditioners or, in arid climates, evaporative coolers (once commonly referred to as *swamp coolers*).

If your home is equipped with an air conditioner or evaporative cooler, you should notice a dramatic decline in usage and electrical bills after implementing the suggestions in this chapter and my book, *Green Home Improvement*. Don't rest on your laurels. There are additional ways to cut the use of these devices, for example, by improving their operating efficiency. For central air conditioners, you can begin by replacing dirty filters. This simple, cost-effective step will make the units work more efficiently and cut energy bills.

If your air conditioner is located in a sunny location, find a way to shade it. Air conditioners work a lot more efficiently if they're not baking in the hot sun all day long. Even an artificial shade structure over an air conditioner will help it function more efficiently — and will cut down on energy use. Just be sure you don't block air flow from the unit.

Next, be sure that the air ducts that distribute cool air in your central air conditioning system are well sealed. Poorly sealed ducts can leak a considerable amount of cool air into unconditioned spaces, for example, attics. Leaks of this nature can easily waste 25 to 35 percent of the energy your air conditioner uses to keep your home cool. Leaks dramatically increase run time and overall energy use.

To seal ducts, use mastic. It is a gooey, long-lasting adhesive that's painted on with a paint brush. Do not use duct tape or aluminized duct tape. Mastic will last much longer than either of these products.

You may also want to insulate cooling ducts that run through unconditioned space. Most local hardware stores carry a couple of different products for insulating large ducts.

Finally, you may also want to call your local utility company or contact a local air conditioning specialist to take a look at your air conditioner to make sure that it is running efficiently.

If you need to replace your air conditioner, purchase a super-energy-efficient model. Don't call around to local suppliers or installers, though. Look first at energystar.gov for a list of the most efficient air conditioners on the market today.

On a related issue, be sure that your new air conditioner isn't oversized. Oversized units are much less efficient because they turn on and off more frequently than right-sized units. Because air conditioners take a while to reach peak efficiency, switching on and off means they run less efficiently over the long haul. They also may not remove as much moisture from the interior of a building.

Be certain the supplier knows all you have done to reduce the cooling load in your home. Give him or her a list of your energy upgrades so he or she can size the system correctly. The energy efficiency improvements you've made should reduce the size of the air conditioner, easily paying off most, if not all, of the costs of your energy upgrade.

Alternatively, you may want to consider installing an energy-efficient heat pump "air conditioner." One common type used largely in warmer or moderate climates like that of the southern United States is the air-source heat pump, discussed in Chapter 4. Air-source heat pumps suck heat out of our homes in the summer and release it into the outside environment. They are generally much more energy efficient than air conditioners. In fact, they are used in a lot of new homes right now.

As you may recall from Chapter 4, heat pumps can also be used for wintertime heating. During the winter, a heat pump extracts heat from the outside air. Using refrigeration technology, the heat pump concentrates the heat and then transfers it into the home.

Evaporative coolers are also widely used in many areas, notably in hot dry regions, to cool homes, stores, restaurants,

Size Matters

Contrary to common belief, an oversized air conditioner is extremely inefficient. It cycles on and off very frequently, which is inherently inefficient, and in so doing, doesn't effectively dehumidify the air, either.

shops, and other buildings. Evaporative coolers draw dry outside air through a wet, porous mat, kept moist by water dripping down from the top of the mat. Dry air passing through the mat evaporates the moisture and is therefore cooled. (Remember, evaporation sucks heat out of air.) The cooled air is then blown into the building, providing economical comfort on the hottest of days.

Evaporative coolers use much less energy than air conditioners, but as noted above are only appropriate in dry climates. Be sure to select the most water- and energy-efficient design available.

Cooling People

The fourth tool in the passive cooling toolbox falls under the category of cooling people directly.

One of the most economical ways to cool people off is turning on a fan. Any fan will do as long as it creates air movement that strips heat from our bodies. You can test this idea by fanning your face with a piece of paper or this book. You'll feel immediately cooler even though the air temperature has not changed. Why does this make you feel cooler?

Heat tends to accumulate in a zone immediately surrounding our bodies called the *boundary layer*. Pushing that warm air aside immediately makes a person feel cooler.

Operating a fan, such as a ceiling fan, has the same effect as lowering room temperature by 4°F (2°C), provided you can feel the moving air. If you can't, it won't help. In such cases, a box fan or an oscillating fan on a nearby table will prove more effective.

Fans use *much* less energy than air conditioners or evaporative coolers. They can also be used in conjunction with air conditioners or evaporative coolers, and are especially useful in hot, muggy climates. Using a fan in conjunction with an air conditioner allows you to raise thermostat temperature settings several degrees, saving considerable amounts of energy. According to the authors of the *Consumer Guide to Solar Energy*, each degree of increase in your thermostat setting cuts cooling costs by about 8 percent.

The Trickiest Climates

Passive cooling is a pretty straightforward process that requires incremental application of many small, seemingly inconsequential measures. But collectively these measures make huge dents in your cooling bills (see Tables 7-1 and 7-2). The hardest of all climates to cool in, though, are hot, humid ones. In a hot, humid climate, heat can't escape from the Earth's surface — it is trapped by atmospheric moisture. So, hot, humid days turn into hot, humid nights.

In hot, humid regions, passive cooling measures often work well early and late in the cooling season when combined with natural ventilation (to purge heat from homes) and fans (to cool people). During the hottest and most humid days of the year, however, passive cooling will likely require assistance from a ceiling fan or two, and perhaps even an air conditioner. If interior daytime temperatures are sufficiently comfortable, an air conditioner can be used to purge heat at night. If that doesn't work, you'll need to run the air conditioner during the day, but run time should be dramatically reduced if you have incorporated the passive cooling measures described in this chapter.

For the most effective cooling, you should monitor outside temperature and humidity levels. In Missouri, I've found that we often have a few hot, humid days in a row, followed by hot, drier days. If the nighttime temperature drops below 75°F, I use the whole house fan to cool the place down. Unfortunately, most people regulate temperature via the thermostat, and pay very little attention to the weather.

BUILDING A BETTER FUTURE

Table 7-2 lists many of the passive cooling measures you can take. By conscientiously applying these measures, you can achieve greater energy independence — and all the positive things that come with that.

Table 7-2 Steps to Passive Cooling		
Strategy	Ideas Presented in Chapter	Ideas of Your Own
Reduce Internal Heat Gain	Use lights sparingly. Turn lights off when not in use. Remove light bulbs in areas where they're not needed to avoid overlighting. Turn water heater temperature down to 120°F (49°C). Install water heater insulation blanket. Insulate hot water pipes. Eat more cold meals in the summer. Cook outside. Use the microwave in the summer. Bake at night. Run exhaust fan when cooking. Use the cold or warm water settings on washing machine. Wash clothes at night. Hang clothes on outside line. Dry larger loads. Close off utility room. Open window to utility room when the clothes dryer is in use during summer. Turn computers and other electronic devices off when not in use. Watch TV more sparingly. Unplug TV and stereo when not in use. Plug TV and stereo into power strip and turn off when not in use. Turn off furnace pilot light during the cooling season. Let pets spend more time outside in the summer. Spend more time outdoors on porches and patios. ☞	

Strategy	Ideas Presented in Chapter	Ideas of Your Own
Reduce Internal Heat Gain	Take shorter showers. Open window when showering. Run exhaust fan when showering. Install an efficient showerhead. Hand-wash dishes. Switch off drying option on dishwasher.	
Reduce External Heat Gain	Plant shade trees. Build artificial shade structures such as arbors and trellises. Install awnings. Install and use window shades. Seal cracks in building envelope. Upgrade insulation. Replace energy-inefficient windows. Repaint with a lighter color. Replace roof shingles with lighter-colored ones or metal roofing or Spanish tiles. Install radiant barriers.	
Purge Heat	Use natural ventilation early and late in cooling season and as much as possible during the height of the cooling season, if your climate permits. Purge heat at night in dry climates. Install and use window fans. Install attic fan. Install whole house fan. Improve efficiency of air conditioning system (seal ducts, replace dirty filters, shade air conditioner, etc.). Replace inefficient air conditioners with more efficient models or evaporative coolers if you are in a dry climate. Install an air-source heat pump.	
Cool People	Use fans. Install and use ceiling fans.	

CHAPTER 8

SOLAR ELECTRICITY

POWERING YOUR HOME WITH SOLAR ENERGY

Solar electric systems are becoming extremely popular in the United States and other developed countries (Germany and Japan, in particular). The recent surge in their popularity is due to several factors, among them financial incentives and forward-thinking political policies designed to create a sustainable energy future.

Solar electricity is surprisingly popular in less developed countries. Although solar electricity is one of the most expensive sources of electricity known to humankind, in rural areas in less developed countries, it's often much cheaper to install solar electric systems in small villages than to string miles of electric lines from central power plants to remote areas.

Solar electricity is also much cheaper than conventional electricity in certain applications in more developed countries. One example is the remote emergency solar-powered call box, a technology that is popping up along highways all across the United States (Figure 1-7). We're even seeing them in cross-walks, where they power lights that help you safely cross the street (Figure 8-1). Why? It is far cheaper to provide solar electricity to these remote installations than to run an electric line to them.

Solar's rural benefit is also evidenced by remote highway warning lights. Flashing lights on a remote Missouri highway, for example, warn drivers of a dangerous intersection (Figure 1-6). Solar electric lights are also found on buoys in major rivers, like the St. Lawrence River.

Solar electricity is often much cheaper

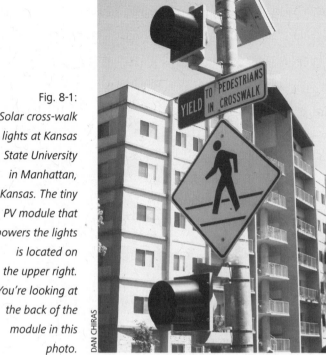

Fig. 8-1:
Solar cross-walk lights at Kansas State University in Manhattan, Kansas. The tiny PV module that powers the lights is located on the upper right. You're looking at the back of the module in this photo.

than conventional power when building a home more than a couple of tenths of a mile away from an electric line — especially if the home is energy efficient. Although some local utilities have generous line-extension policies, many charge $20,000 or so to run a line a few tenths of a mile — and upwards of $50,000 to run a line a half mile. That $50,000 investment in electric poles and electric wire to hook up your home to the grid could be used to purchase a large solar electric system, possibly one even double the size of what an energy-efficient family would

need (especially after subtracting financial incentives). One of my students once spent $60,000 to run electric lines to power two tiny cabins on land she owned in southern Colorado. She was surprised to learn that a $10,000 solar electric system would have sufficed.

Bear in mind that the fee a utility charges to connect a home or business to their network doesn't pay for a single penny's worth of electricity. It just buys you utility poles, an electric line, a meter, and a connection to the grid. What is more, you may need to pay the connection fee upfront, although some utilities will amortize the cost of connection, adding $200 to your electric bill for the next 20 years. In contrast, a $25,000 to $50,000 investment in an off-grid solar electric system will provide you and your family a lifetime of electricity (although you will need to replace batteries every 7 to 15 years, depending on how well you care for them).

Solar Electric System Incentives

Solar electric systems can make sense in places where electrical costs are extremely high. In California, Hawaii, and Germany, for instance, conventionally produced electricity goes for premium prices.

Solar electric systems also often make economic sense when financial incentives are available to utility customers. Financial incentives come in several forms. The most common are federal tax credits and local utility rebates. Together, they can decrease the initial cost of a solar electricity system by about one half. US businesses can receive their federal tax credit almost immediately by filing for a grant from the US Treasury Department once the system is installed. There's no need to wait for tax season. And businesses don't even need to have a tax liability to qualify for a tax credit.

US businesses can benefit from an additional economic incentive: accelerated depreciation on federal income tax. This allows businesses to depreciate the cost of a solar electricity system in five year's time, which effectively reduces the initial cost of the system by approximately 30 percent, depending on the business's tax bracket.

In the United States, the US Department of Agriculture offers grants for renewable energy systems that cover 25 percent of cost of the system. These grants are available to businesses in rural areas (areas with a population under 50,000).

Generous incentives such as these dramatically reduce the initial cost of solar electric systems and the cost per kilowatt-hour of electricity they generate over their lifetime, often making the electricity cost competitive with conventional electricity.

States and utilities also provide incentives. The state of Illinois, for instance, currently offers generous incentives totaling about 50 percent of the cost of a solar electric system. New York State offers its residents a 25 percent tax credit. The state of Florida currently exempts solar electric systems from certain taxes, and some areas like Colorado's Roaring Fork Valley, home to the prosperous town of Aspen, offer zero-interest loans for homeowners who install solar electric systems. A number of states, including Colorado and Missouri, require investor-owned utilities (IOUs) to provide rebates to customers who install solar electric systems. The rebates are frequently about $2 per watt of installed capacity (discussed shortly). To put this in perspective, it typically costs about $6 to $8 per watt to install a grid-connected system.

And there is more good news. In 2009, the price of solar electric modules tumbled and has remained low. (They've never been cheaper.) Thus, if you live in a state like Arizona, Illinois, New York, Colorado, or New Jersey, or are served by a utility that offers generous financial incentives, you may want to retrofit your home or business right now. If you own a business, especially one in a rural area, you can acquire additional benefits like a USDA grant for 25 percent of the system cost.

To determine whether you are qualified to receive a rebate, call your local utility. Better yet, check out incentives on the Database for State Incentives for Renewable Energy (dsireusa.org).

If there aren't any incentives available in your area, you can still consider installing a solar electric system as a hedge against rising prices. Or you may, like me, want to install a system to achieve energy independence and reduce your environmental impact. Because solar electric systems can be expanded, you can invest in a few modules at a time and build your system over a period of five to ten years. For more on financing a renewable energy system, see Allan Sindelar and Phil Campbell-Graves's article on the subject in *Home Power* magazine, Issue 103.

Fig. 8-2: *Simplified anatomy of a grid-connected solar electric system.*

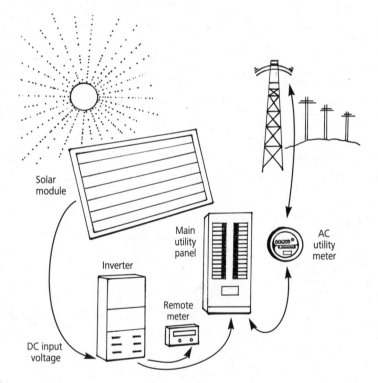

WHAT IS A SOLAR ELECTRIC SYSTEM?

Solar electric systems fit into three main categories: grid-connected, grid-connected with battery backup, and off-grid. The differences will become clear shortly. Figures 8.6 to 8.8 show the three options and their main components.

Grid-Connected PV Systems

Grid-connected solar electric systems are the simplest of the three options. As shown in Figure 8-2, they consist of three main parts: solar modules, an inverter, and the main service panel or breaker box. Let's begin with the solar array.

Like all other solar systems, the grid-connected solar electric system consists of one or more solar electric modules also known as *photovoltaic* or PV *modules.* ("Photo" refers to light, and "voltaic" means electricity.) Each solar module consists of numerous smaller solar cells, usually round, square, or rectangular cells that are about 1/100th of an inch thick and made from silicon (Figure 8-3a). Silicon comes from two highly ubiquitous materials, high-grade sand and quartz rock — both of which are made of silicon dioxide. Silicon is extracted from these materials to produce polysilicon.

The solar cells most widely used today consist of two thin layers of silicon. As illustrated in Figure 8-3b, each cell in a module has numerous thin metal electrical connections (metal contacts) on the surface that gather up electricity generated by the cell.

A solar or photovoltaic cell is a solid-state device that absorbs visible light and converts its energy to electricity. How a PV cell works is fascinating, but beyond the scope of this book. Suffice it to say that when sunlight strikes a PV cell, it liberates electrons from silicon atoms. Thanks to some ingenious chemistry (boron and phosphorus are added to different layers of the PV cell), these electrons are forced to migrate to the surface of the cell, where they are drawn away by the metal contacts on the cell's surface, as shown in Figure 8-3b. Flowing electrons form an electrical current.

Because silicon reflects about 35 percent of the light striking it, the cells are coated with a thin, anti-reflective layer of silicon monoxide or titanium dioxide. The back of the cell consists of a thin layer of metal that completes the electrical circuit, as shown in Figure 8-3b.

Fig. 8-3: *(a) Two solar cells. (b) Note that solar cells consist of two layers. Light striking the layers liberates electrons from the silicon atoms. These atoms flow from the deeper layer to the more superficial layer and then are drawn off by the metal contacts on the surface, creating direct current electricity.*

Silicon cells are mounted in an aluminum frame, the module casing, and wired in series to boost the voltage. (Wiring in series increases the voltage.) The unit is sealed by a clear layer of glass, which prevents moisture from reaching the cells and also resists the pounding force of hail stones. (The glass in PV modules can withstand the force generated by a ball bearing striking it at 120 miles per hour.) Typical modules produce between 40 and 250 watts of electrical power under peak sunlight conditions. In most new systems, installers install 175 to 250-watt modules. Some manufacturers are producing even larger modules that produce around 300 watts.

Numerous modules are wired together to form a *solar array*. The array includes the modules plus a rack on which the modules are mounted.

PV modules can be mounted on buildings or on the ground. In urban and suburban areas, they are typically mounted on a durable metal rack on the roofs of homes. In rural areas, PV modules are mounted on roofs or on the ground. Whatever location you choose, be sure it has full access to the sun year round for optimal performance.

In some installations, the modules are mounted on a rack attached to a pole. Some pole-mounted systems include a tracking device that enables the array to follow the sun from sunrise to sunset each day. These are known as *single-axis trackers*. Other, more elaborate trackers follow the sun from sunrise to sunset, and also adjust the tilt angle of the array to accommodate the change in the sun's angle. (The sun is higher in the sky in the middle of the day and in the summer.) These are known as *dual-axis trackers*. (Be sure to install your solar electric modules so they are out of the reach of vandals or livestock.)

Tracking arrays boost the output of a PV system by 10 to 40 percent. For best performance, a tracking array should be located in an open field or large, open (treeless) lawn that permits access to the sun from sunrise to sunset. Trackers perform better in northern regions because they experience longer days in the summer. Dual-axis trackers perform only slightly better than single-axis trackers.

Trackers add to the cost of a PV system, but the higher cost can be offset by the value of the electricity produced. Because trackers have controls, motors and other moving parts, they may require periodic repair or replacement of damaged parts, such as the controller that runs the motors. Because of this, you may be better off adding a few more solar modules to a fixed array (one that doesn't track the sun) rather than installing a tracker. Be sure to consider the cost of a tracker and how it affects the cost per kilowatt-hour of electricity the array will produce.

Solar modules produce direct current (DC) electricity when the cells are struck by sunlight. Direct current electricity consists of electrons flowing in wires in a single, fixed direction. The energy they contain can be used to power motors, lights, home appliances, and a host of other electronic devices.

The DC electricity produced by a solar array is carried away by wires that lead into the home. As you may know, however, virtually all of our homes use another type of electricity, known as *alternating current* (AC). AC electricity consists of electrons that cycle very rapidly back and forth through a wire. In North America, standard household current is 60 hertz, or 60 cycles per second. That means that in standard household electrical current, the electrons cycle back and forth 60 times per second. The energy these electrons carry is used to power a wide range of 120- and 240-volt appliances and electronic devices in our homes and businesses.

In order for the DC current produced by PV cells to power our homes, then, it must be converted into AC current. This task is relegated to an *inverter*, shown in Figure 8-4. The inverter converts DC electricity into AC electricity. Inverters also change the voltage of the incoming DC electricity to 120 volts or 240 volts (needed for some appliances, like electric clothes dryers). Inverters are generally

Fig. 8-4:
This quiet, efficient inverter by Sunny Boy converts direct current produced by PVs (or a wind generator) into alternating current and boosts the voltage to 120. Inverters that are used in systems with batteries typically contain battery chargers.

Fig. 8-5:
The Enphase microinverter. This inverter is mounted on the rack behind each module. Modules plug into the inverter, which produces 240-volt alternating current electricity.

located near the main panel or breaker box.

In recent years, Enphase has introduced a promising new product known as the *microinverter* (Figure 8-5). Interestingly, microinverters are attached to each module in a PV array. They are only used in grid-connected systems. While traditional inverters produce either 120-volt or 240-volt AC current that travels to the main electrical service box of your home, microinverters produce 240-volt AC electricity.

AC Devices Run on DC

Even though we plug all electronic devices and appliances into wall sockets that deliver alternating current, many of them actually run on DC current. The small black transformer of a laptop computer, for instance, converts the AC to DC and decreases the voltage. Even TVs and stereo equipment convert incoming AC to DC.

They are wired directly into the main panel via a circuit breaker.

The breaker box contains electrical circuits, known as *branch circuits*, that supply various parts of a home or business. It also contains circuit breakers to protect the wires in a home or business from overcurrent — carrying more electricity than they are rated to carry. Overcurrent can cause wires to heat up and start a fire.

In grid-connected systems, electricity from the solar array powers electronic devices in our homes and businesses. Surplus electricity is automatically sent into the electrical grid (Figure 8-6a). Surplus power simply flows out through the main service panel through the utility meter. It then flows through the electrical wires that connect to the local utility network. These are the wires that connect homes and businesses to regional power plants; they are often referred to as the "electrical grid." (Technically, the electrical grid consists of the high-voltage electric lines that

crisscross countries. The local electric lines are the utility network.) The surplus sent to the electrical grid is consumed by your neighbors.

In many grid-connected solar electric systems, electricity flowing to the grid runs through the same meter that keeps track of electricity delivered to your home from the grid. Electricity you are sending to the grid, however, runs the meter backwards, so you are credited for the electricity you are supplying to the grid. (More on this shortly.) In other systems, electricity flows through a separate meter that keeps track of how much electricity you supply to the grid. Utility meters are fairly large, tamper-proof glass-encased meters installed by the utility company. They keep track of monthly energy production and consumption so the utility can bill you at the end of each month.

Additional Components: Meters and Disconnects

All code-compliant solar systems include two additional components: meters and various disconnect switches. Meters are typically located on the inverter and are used to measure electrical production by a PV system. They'll tell you how much electricity the PV system is producing and other important information.

Meters typically give readings on the watts, amps, and volts of incoming DC electricity and the same for outgoing AC — AC produced by the inverter. They keep track of daily production and long-term production in kilowatt-hours. (Which is how we know that in its first year, The Evergreen Institute's 5-kW PV system produced over 6,200 kilowatt-hours of electricity!). If an inverter doesn't come with a built-in meter, you can buy a separate unit and connect it to your system, and it will provide the data.

Grid-connected solar electric systems also contain a couple of safety switches, called *disconnects*. Disconnects enable homeowners or service personnel to shut power down to prevent electrical shock when working on the system.

As shown in Figure 8-6b, a DC disconnect is located between the solar array and the inverter. It is used to terminate the flow of electricity to the inverter. Today, modern inverters have a DC disconnect built into them, so it is not always necessary to install a separate DC disconnect.

The other disconnect, the utility-accessible AC disconnect, is located between the inverter and the main service panel. It is used to shut off the flow of electricity from the inverter to the household circuits and the grid. Utility workers use the AC disconnect to shut off the system if the lines go down and they need to work on the electric lines to your home or in the neighborhood. However, because inverters shut off automatically when they sense that the grid is down, an AC disconnect is not really needed; your utility may recognize this and not require you to have one.

How Utilities Meter Electricity in Grid-Connected Systems

Unlike off-grid systems, grid-connected PV systems have no physical on-site electrical storage capacity. That is, they have no means of storing electricity for later use, for example, at night when the PV modules are inactive. Solar homeowners, however, use the electrical grid as their "storage battery." That is to say, the electrical grid accepts excess electricity when a solar electric system is producing more electricity than a home is using. Excess electricity that's transferred onto the grid is used by one's neighbors. However, at night, when a system is no longer producing electricity — or during the day when a home requires more electricity than the PV system is producing — electricity is drawn from the grid.

To keep track of this, many utilities simplify matters by installing one electrical meter. Older meters contain a flat disk that runs forward when electricity is being drawn from the grid. Each rotation of the disk represents a certain amount of electricity consumed by a household. Rotations of the disk are converted to kilowatt-hours by the meter.

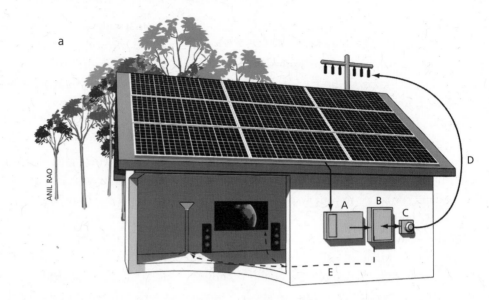

A Inverter
B Breaker box
   (main service panel)
C Utility meter
D Wire to utility line
E Circuits to household
   loads

Fig. 8-6:
*(a) This schematic
shows the major
components of a
grid-connected
solar electric system
and illustrates how
surplus electricity
is backfed onto
the electric grid.
(b) This diagram
shows more detail,
including charge
controllers and
disconnects.*

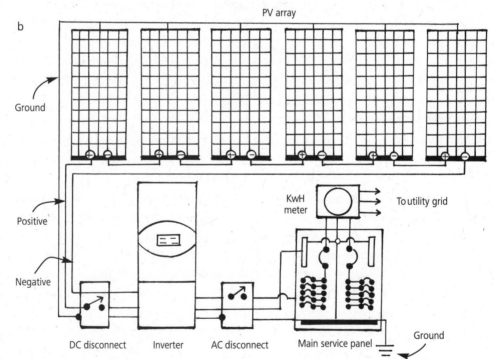

When electricity is flowing on to the grid, the meter runs backwards, subtracting from the amount of electricity your home has consumed previously. The kilowatt-hour reading on the meter decreases.

All utilities in the United States are required by federal law to accept electricity from small generators. Most are required by state law to engage in *net metering*. In very basic terms, what net metering means is that every bit of electricity you feed onto the electric grid can be withdrawn at no charge during the billing cycle.

Two types of net metering are in place in the United States: monthly and annual. If you are on a monthly net metering plan, surpluses generated during the day can be withdrawn from the grid whenever they are needed, at no charge. If you are on an annual net metering plan, these surpluses can be carried from one month to the next. So, if a PV system generates a surplus in September, it can be used in October or even November or December. Annual net metering systems are a lot like cell phone plans that allow customers to roll minutes over from one month to the next for up to one year.

To see how these work, let's start with a monthly net metering plan. Suppose your system delivered 400 kilowatt-hours of electricity to the grid but your household consumed 400 kilowatt-hours at night and at other times. If this were the case,

your cost for electricity would be zero — although utilities usually charge you a fee (around $10–$20 per month) to read the meter.

If your PV system produced 200 kilowatt-hours less than you consumed, you would be charged for the consumption. But what would happen if your system produced a surplus during the monthly billing cycle — say 400 kilowatt-hours more electricity than you withdrew from the grid? This is called *net excess generation*.

Most states in the United States now have net metering laws that dictate how utilities must deal with surpluses. Basically, there are three options. First, utilities can be granted the surplus. That is, they take your 400 kWh, sell it, and made a nice profit. Second, they can be required to pay you their *avoided cost*, that is, what it costs the utility to buy or generate electricity. Avoided cost, or wholesale rate, is typically about one fourth of the retail rate. If the utility is selling electricity to its customers for 10 cents per kWh, the avoided cost is typically about 2.5 to 3 cents per kWh. So, if your system produced 400 kWh

More on Net Metering

To view some frequently asked questions on net metering, you can visit the website sponsored by *Home Power* magazine, a leader in home renewable energy: homepower.com.

of surplus, you'd receive a check or, more likely, be credited on your bill for the 400 kWh at 2.5 to 3 cents per kWh. In other words, you'd be paid or receive a deduction from your bill (e.g., monthly service charge) of $10 for the $40 worth of electricity you delivered to the grid (which they sold for 10 cents per kWh).

The third option is that the utility is required to reimburse customers for net excess generation at the retail rate—the amount they charge their customers. This is, of course, the best deal.

In states with annual net metering, surpluses generated in sunnier months accumulate and can be used in less sunny periods. At the end of the one-year period, the utility may keep the surplus, pay for it at avoided cost, pay for it at retail rates, or continue to roll over excesses into ensuing years — it all depends on state law. Kentucky allows surpluses to be carried indefinitely. In Colorado, customers of investor-owned utilities can opt for indefinite rollover.

As you study this issue, you will find that there are three types of utilities: (1) large, investor-owned utilities that typically serve urban areas; (2) municipal utilities, that is, power companies owned by cities and towns; and (3) rural cooperatives, nonprofit organizations that typically buy electricity from various producers and then sell it to their rural customers. Different rules apply to each of them. For instance, in Colorado, investor-owned utilities are required to pay customers wholesale rates at the end of the year, but rural co-ops and municipal utilities — called "munis" — can pay whatever they deem reasonable. Unfortunately, many rural cooperatives have been antagonistic toward solar and wind systems, and may not be easy to work with. However, I've worked with two rural cooperatives in Missouri and found them both to be very cooperative.

As of September 2010, 43 states, Puerto Rico, and the District of Columbia had mandatory net metering laws in place. Texas and Idaho initiated voluntary programs. South Dakota, Tennessee, Mississippi, and Alabama have none. See the sidebar, "Does Your State Net Meter?" for a website that will help you determine the policy in your state.

Does Your State Net Meter?

To see if your state offers net metering and to learn the specifics of your state's program, visit the US DOE's website at eere.energy.gov/ and search for "net metering."

Grid-Connected Systems with Battery Backup

Some homeowners elect to install batteries in their grid-connected systems to provide backup power in case grid power

a

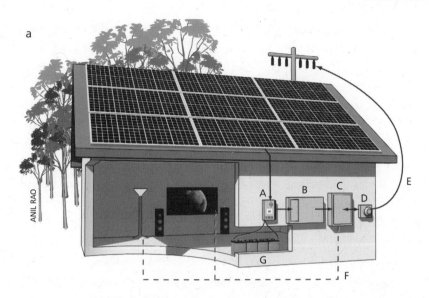

A Charge controller

B Inverter

C Breaker box (main service panel)

D Utility meter

E Wire to utility service

F Circuits to household loads

G Back up battery bank

ANIL RAO

b

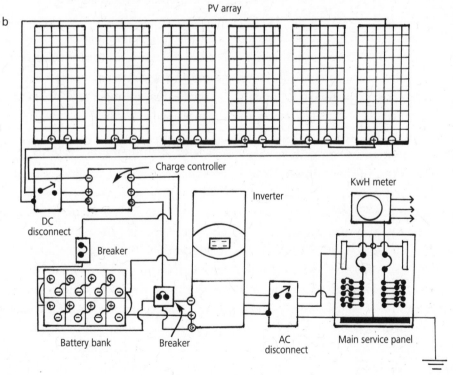

Fig. 8-7:
(a) This schematic shows the major components of a grid-connected solar electric system with a battery bank to store backup electricity. (b) This diagram shows more details, including controllers and disconnects whose functions are explained in the text.

goes down (for example, if a tree branch falls on a utility line and shuts down the electrical power system supplying you and 65 million other people). A typical grid-connected system with batteries is shown in Figures 8-7a and b.

In grid-connected systems equipped with batteries, the solar array produces electricity that keeps the batteries fully charged at all times. Solar electricity also feeds live circuits in the house during the day. When the batteries are full and all demand is met inside a home or business, excess electricity is backfed onto the electrical grid. At night, power required for household use is delivered by the grid. The grid also supplies electricity when demand in a home or business exceeds the output of the PV array. The batteries are called into action *only* when grid power fails, although a homeowner can switch to battery operation if desired via a manual transfer switch.

A grid-connected system with battery backup includes all of the components of a grid-connected system, including disconnects for safety. Grid-connected solar electric systems with battery backup, however, also require a couple of additional components, including the batteries, and a component we haven't discussed yet — the charge controller — as shown in the schematic in Figure 8-7b.

Charge controllers are devices that regulate the charging of the batteries during normal operation and during times when the PV system goes off grid — for example, when the electric utility experiences a blackout and the system switches to battery power. Charge controllers not only regulate battery charging, they protect batteries from overcharging (how they do this is beyond the scope of this book). Overcharging can damage the lead plates inside batteries, reducing battery life. (If you want to learn more, check out my book, *Power from the Sun.*) The charge controller is housed in a metal box that is mounted on the wall near the inverter and the batteries.

Grid-connected systems with battery backup also contain meters to monitor electrical production, battery voltage, and battery storage — how much electricity is stored in the batteries. Some meters are on the charge controller and others are on the inverter.

Off-Grid (Stand-Alone) Systems

Next on the list is the off-grid system, also known as a *stand-alone system.* Figures 8-8a and b show the anatomy of an off-grid solar electric system. As you can see, it is very similar to the grid-connected system with battery backup. You will notice two distinct differences, however. First, there is no connection to the grid. Thus, no surplus electricity can be backfed onto the grid. This system produces electricity that meets immediate demand but also

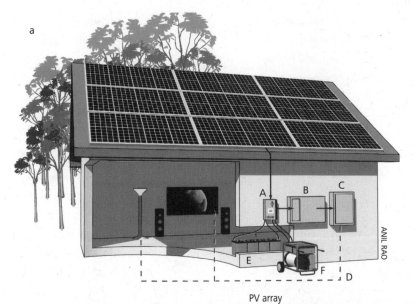

a

A Charge controller
B Inverter
C Breaker box
D Circuits to household loads
E Battery bank
F Back up generator

ANIL RAO

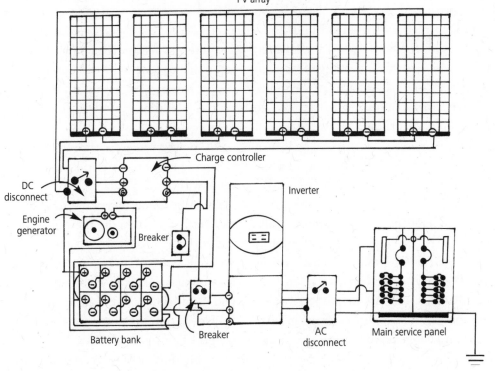

PV array

Charge controller

DC
disconnect

Engine
generator

Breaker

Inverter

Battery bank

Breaker

AC
disconnect

Main service panel

Fig. 8-8: These
schematics shows
all the components
of an off-grid solar
electric system,
including charge
controllers and
disconnects.

stores electricity in batteries for use at night or during long, cloudy periods. The batteries also supply electricity when the immediate demand exceeds the output of the PV array. Second, this system usually includes a backup generator that can be run to charge the batteries when they run low. Some people install wind turbines to serve as a second charging source.

As illustrated in Figure 8-8, the main power source of an off-grid system is the PV array. Electricity flowing from the array into the house first travels to the charge controller. It can then travel to the batteries and to the inverter, where it is converted to AC electricity. At night, all electricity required in an off-grid home comes from the batteries via the inverter, which converts the low-voltage DC solar electricity to high-voltage AC current. (Batteries are typically wired at 12, 24, or 48 volts, depending on the electrical demand of a home.) The electricity is then delivered to open (active) circuits in the house via the main panel, a.k.a. the breaker box.

What happens if the batteries are full?

In off-grid systems, the charge controller terminates the flow of electricity from the PV array to the batteries when they are full. This prevents the batteries from overcharging. This task is managed by the high-voltage disconnect in the charge controller.

Like other solar electric systems, off-grid systems require an AC and DC disconnect to enable a repairman to service a system without being electrocuted.

Off-grid solar electric systems like the one shown in Figure 8-8 also include a low-voltage disconnect housed in the inverter. When it senses low battery voltage, it terminates the flow of electricity from the batteries to the inverter. This prevents over-discharging — that is, draining the batteries too low. Over-discharging can permanently damage a battery bank.

Power Centers

Off-grid solar electric systems can seem pretty complex; they have enough connections to frustrate the far more complex neuronal pathways in the human brain. There is, however, a way to avoid some of the confusion: install a power center (Figure 8-9).

A power center is the grand central station of off-grid and grid-connected systems with battery backup. Power centers contain the connection points to which the electrical wires from the charge controller, the inverter, the batteries, and the solar array all connect. The power center houses the circuit breakers that protect the wires and the disconnects that allow servicing.

I loved the Ananda power center in my off-grid home in Colorado. In the classroom building at The Evergreen Institute, my students and I installed an Outback (Flexware 500) power center for the grid-connected system with battery backup. Unfortunately, the Flexware 500 was not pre-wired and came with rather generic wiring diagrams. The frustration mounted as we started to wire the system, but faded quickly once an experienced installer in our area handed us a good wiring diagram. So, be sure to get a power center that is pre-wired — or ask for a wiring diagram that matches your application. I strongly recommend that you consider a power center if you are going to install a PV system with batteries. It will cost a bit more than buying the components separately, but it is well worth it when you factor in the massive reduction in cerebral hemorrhaging that will result from attempts to buy all of the components separately and

wire them in some meaningful and functional manner (that meets all of the safety requirements of the National Electrical Code).

Another good reason to consider buying a power center is that they are very popular among electrical inspectors — who typically know very little about solar electric systems and are often totally confused by battery-based PV systems. When an inspector sees a power center, he or she knows that all of the essential safety circuitry and all of the vital components required in a solar electric system are present and accounted for, housed in a single assembly. Rather than have to go through the system with a fine-toothed comb, they'll take one look at your system, find the UL (Underwriter's Laboratory) sticker, and then check it off on their

Fig. 8-9:
For stand-alone systems, the power center is a helpful organizer of components. It contains meters, charge controllers, and disconnects.

OUTBACK

inspection sheet and move on to their next inspection.

DC Circuits

So far, the discussion of solar electric systems has focused on converting the DC power they produce into the AC power used in our homes. Solar electric systems can also be designed to use DC power directly — for all circuits, or just a few. In my home in Colorado, for instance, I ran a small, 24-volt DC electric pump to pump water from my cistern to the house. I did this for one simple reason: to bypass the inverter. By doing this, I save energy and simplify the system.

How does bypassing the inverter save energy?

Inverters consume energy when they're operating. For example, when my inverter converts 24-volt DC electricity from my batteries to 120-volt AC current, it loses about 10 percent of the energy that goes through it. By bypassing the inverter with a DC circuit directly to the water pump, I raise the efficiency of my system.

Other appliances and consumer electronics can also be run on DC power. For example, SunFrost refrigerators, which are one of the most efficient refrigerators on the market today, can be ordered with DC components at no extra cost. Because refrigerators account for about 8 percent of a family's daily electrical consumption, you can save substantially by purchasing a DC refrigerator and bypassing the inverter.

Ceiling fans can also be ordered in DC models. Here, though, you need to think carefully. I was planning on doing this until I found out (in 1995) that DC ceiling fans cost $200 each, compared to $40 each for standard AC ceiling fans. I couldn't justify spending $480 more for three ceiling fans to bypass the inverter. Many other DC appliances are available, but they tend to cost more and are typically designed for use in RVs or cabins. They're not ideal for household use, unless you've pared down your demands. They also tend not to stand up to daily use very well. Low-voltage DC electricity doesn't travel very efficiently through small wires, so large and more expensive wires are required, which adds to the cost. DC wiring requires special, more costly receptacles. All in all, the electricity you save by bypassing the inverter may be lost by transmitting low-voltage DC through your home. As a rule, then, DC circuits aren't of much value in most homes. But don't close the book on them. You may want to install at least one DC circuit to power a lightbulb in the utility room (where the inverter is located) in case of emergency.

System and Battery Maintenance

Before you make up your mind what kind of solar system you want, you should consider system maintenance. A grid-connected system is virtually maintenance-free. If you live in a dry dusty area or a polluted city, you may need to periodically wipe, rinse, or dust off your PV modules from time to time. In most locations, however, rain does the job free of charge. (I've never cleaned mine in Colorado, and we only get about 20 inches of precipitation per year. I've also never cleaned my PV modules in Missouri, where it rains a lot.)

In snowy climates, you may need to sweep snow off your array, being careful not to damage the modules. Even a tiny amount of snow blocking the lower edges of PV modules in an array can reduce production by more than 90 percent. (When snow melts on an array, it tends to slide down and may accumulate along the bottom of the modules.) If you live in a snowy climate, be sure to install PVs within reach so you can remove snow easily.

Off-grid PV systems are more complicated than grid-connected systems and require a lot more maintenance. Why do I typically recommend against them? Because lead acid batteries used in off-grid systems require babying. Batteries function optimally at 70°F (21°C). At higher or lower temperatures, their performance plummets. To ensure optimal performance and long life, you need to house batteries at their preferred temperature — which means in a space where temperatures

Battery Room Safety Equipment

A battery room should be equipped with a smoke detector, an appropriate fire extinguisher, and a case of baking soda to neutralize acid in case of a fire or spill. Rubber gloves and safety glasses are also essential for dealing with batteries.

remain between 50 and 80°F year round. Many people house their batteries in sheds or in garages, often inside a sealed battery box. But take note: the battery room or battery box *must be* vented to the outside so that hydrogen gas produced when batteries are charging can be safely vented outdoors. (Hydrogen is explosive at certain concentrations.) Remember to include these costs in your calculations; building a safe battery box and outfitting a place to keep it costs money.

Watering Your Batteries

Batteries naturally lose water when being charged, either by electricity from the PV modules or the generator. To keep the lead plates from drying out, which can ruin them, you need to periodically refill your batteries, one cell at a time, with distilled or de-ionized water. Batteries should be checked every month and topped off if necessary (be sure not to overfill), according to Richard Perez, who publishes *Home Power* magazine. Refilling takes a half hour or so, depending on the size of your

Fig. 8-10: *These Hydrocaps replace the standard caps that come with lead acid batteries. A catalyst in the reservoir captures hydrogen and oxygen given off by the breakdown of water and recombines them, reducing water loss by 90 percent, thus cutting down on routine maintenance.*

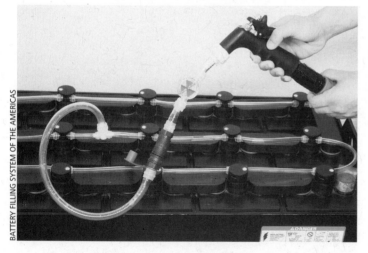

Fig. 8-11: *This battery filling system is designed for those who don't want to have to top off their batteries every three or four months. Float valves in the caps tell when a cell needs replenishment. Distilled water flows into the cells from a central reservoir (not shown here).*

battery bank and how accessible the batteries are.

To reduce water loss and battery maintenance, two companies have produced special battery caps: Hydrocaps and Water Miser caps. Both caps are designed to replace standard cell caps that come with lead acid batteries.

Hydrocaps contain a small catalyst-filled reservoir (Figure 8-10). The platinum catalyst converts the hydrogen and oxygen released by charging batteries into water (see sidebar, "Reducing Battery Maintenance," for explanation). Hydrocaps reduce water losses by around 90 percent. (As noted in the sidebar, you can also install automatic and manual battery filling systems to reduce battery maintenance.)

Water Miser caps also reduce water loss, but only by 30 to 75 percent. They also function differently. Rather than trapping hydrogen and oxygen and converting it back to water, the Water Miser caps trap water vapor and the fine acid mist that evaporates from the battery fluid. The liquid water and acid then drip back into the batteries. Of the two, I prefer Hydrocaps. They cost more than Water Miser caps but do a better job at reducing water losses.

Another way to reduce maintenance is to install a battery filling system, shown in Figure 8-11. Battery filling systems automatically fill batteries from a central reservoir filled with distilled water, totally eliminating this time-consuming chore. Battery filling systems consist of a series of plastic tubes connected to a central reservoir in the battery room. Each battery cell is fitted with a special cap equipped with a float valve. When the water level drops in a cell (there are usually three cells per battery), the valve opens and allows water to fill the battery to the proper level. Unfortunately, the unit costs about $16 per cell or $48 per battery. A system for a battery bank with 12 batteries would cost about $575. Despite the cost, I installed one on my battery bank in Colorado and found that it reduced battery maintenance (filling) from 30 minutes to 5 minutes, making the investment well worth the money.

Equalizing Your Batteries

Batteries also need to be *equalized* from time to time — usually, every three or four months. Equalization is a controlled overcharge of the batteries.

Batteries are typically equalized by running an external gas-powered generator (commonly called a "gen-set"). The process is pretty simple: to equalize batteries, you start up the generator, which is connected to a battery charger located in the inverter. Then you flip on the circuit breaker in the line connecting the generator to the inverter/battery charger. At this point, the inverter/battery charger takes over. When equalizing batteries, electricity

Reducing Battery Maintenance

When batteries are being charged, electricity passing through water breaks it down into hydrogen and oxygen gases. Hydrogen and oxygen enter the catalyst-filled reservoir in the Hydrocap where they recombine to form water, that then drips back down into the battery. This not only reduces the need for you to add water, it also reduces the danger of hydrogen gas being emitted from battery banks.

is fed into the battery bank very rapidly at first. As the batteries are charged, however, the flow of electricity gradually slows. At the end of the process, electrical current trickles into the batteries. When the batteries are fully charged and equalized, the inverter shuts off, and you're done.

Equalization achieves three important goals. It raises the voltage of each cell to the same level for optimal function, hence the name "equalization." (A battery bank performs only as well as the lowest-voltage cell in the array.) Equalization also cleans the plates in your batteries by driving the lead sulfate off your plates. Lead sulfate forms on the lead plates of batteries when they discharge — that is, give off electricity. Lead sulfate accumulating on the plates inside a battery reduces the amount of electricity a battery can store, greatly hindering battery function. (For those who know a bit about batteries and lead acid battery chemistry: lead sulfate forms on the lead plates that feed into the anode [the negatively charged pole] as well as those that feed into the cathode [the positively charged pole]).

Periodic equalization also helps mix the sulfuric acid in which the lead plates are immersed. All in all, then, equalization helps restore batteries and increases their lifespan.

Are Batteries for You?

Battery-based systems have their advantages, but you should be aware of their disadvantages. They are costly. A single battery bank may contain one to two dozen batteries costing $200 to $300 each. The battery bank needs a home, as noted earlier, that is, some safe place where batteries can be kept warm but not hot. This costs money. And don't forget that your battery bank needs to be wired into your solar electric system, which also requires a bit of money. In addition, batteries require all the periodic maintenance just described.

Optimistically, lead acid batteries only last five to ten years, according to Johnny Weiss, director of Solar Energy International, depending on how you take care of them, and how heavily you draw on them. Replacing batteries costs money.

Which System is Best for You?

You've got three choices in solar electric systems. You can install a grid-connected system, which is by far your cheapest and easiest option. When installing a grid-connected system in a brand new home, you'll have to run electric lines to your home and install a utility meter or two to track electricity. In new and existing homes, you will also probably be required to install an AC disconnect, so fire fighters can shut your solar system off in case of a fire or utility company employees can shut off your system if they're working on the lines in the neighborhood. They don't want electricity flowing onto the grid from your system while they're working on nearby lines for fear of electrocution.

In truth, AC disconnects are really unnecessary, so some building departments no longer require them. Departments that still insist on AC disconnects do not properly understand that all modern inverters (UL-1741 listed inverters) automatically terminate the flow of electricity from a solar electric system to the grid when they sense a drop in line voltage in the grid — either a blackout or brownout.

One serious disadvantage of a grid-connected system, however, is that when the grid goes down, so does your system. It can be the sunniest day of the year. It doesn't matter. Your PV system will stop producing electricity so long as the inverter detects a blackout or brownout. Once these events are over, the inverter will turn back on automatically, and restart energy production.

Your second option is a grid-connected system with a battery bank. Or you can go off-grid entirely, by installing an off-grid system. When consulting with clients on system type, I typically suggest that clients start with a grid-connected PV system. Use the grid as your battery bank and suffer through sporadic power outages with the rest of us. If outages are frequent, long, a major inconvenience, or a health or safety issue, then a grid-connected system with battery backup might be a better way to go.

If you are installing a system on an existing home, be sure to contact the local utility company well in advance and ask for a copy of their interconnection agreement, which you must sign and submit along with system information, most notably a description of the system and the automatic disconnect feature. (As just noted, any inverter with a UL-1741 listing will comply with utility regulations for automatic disconnection. It's called *anti-islanding.*) The form may require an installer's signature, attesting that your system will be designed and installed according to the National Electric Code.

Be sure to apply for a building permit. You may need a permit that covers the structural aspects of the project — rack design and the ability of your roof to support an array — and you will probably need an electrical permit. In some parts of the country, like many counties in rural areas, no permits are needed. You are on your own. Nonetheless, you should be sure your system is properly installed for the safety of everyone involved.

If you are considering installing a solar electric system, you obviously have to think all of this through very carefully. To hold costs down, you might want to begin with a small, grid-connected system. Five 200-watt panels would give you a 1,000-watt system, which would cost around $6,000–$8,000 (including installation), depending on the type of module you buy and the difficulty of installation. Federal tax credits would trim 30 percent off the price tag. Other incentives could further reduce the cost.

A small grid-connected PV contains all the main components of a full-fledged solar electric system. That is, you've got all of the expensive hardware in place. You can add modules later as your finances permit, so long as you have sized your inverter properly. (I recommend oversizing the inverter initially to match possible future demand.) Be sure to install identical modules, too. You can even add a battery bank at a later date, if you want. Though you would need to install a new inverter. And you could eventually disconnect from the electrical grid entirely if you feel comfortable that an off-grid system will work for you.

Efficiency First!

If you decide you are going to join the growing legion of solar electric home-owners, the first thing you need to do is to make your home as efficient in its use of electricity as humanly possible. You'll save a fortune on your solar electric system if you do. As a general rule of thumb, every dollar you invest in efficiency will save you $3 to $5 in system costs, before rebates. Thus, $1,000 invested in measures that cut electrical demand by 25–50 percent can save you $3,000 to $5,000 in system costs. Chapter 2 outlines efficiency measures that will help you reduce electrical demand. While you are at it, be sure to eliminate or reduce phantom (ghost) loads. Although individual phantom loads are small, remember that they're on 24 hours a day. Many tiny loads in a house add up, and servicing them can add substantially to the cost of your system.

Sizing a System

Sizing a solar electric system can be tricky, especially if you are installing a system on a house you're just moving into or a house you are building. In such cases, how do you know how much electricity you'll need?

What most people do is make a list of all the appliances and electronic devices that will be used in the house. Once this is complete, they estimate how long each appliance or device is used each day. They then multiply the power consumption (in watts) of each device. (Power consumption is listed on all electronic devices on a small plate or sticker on the back of the unit. Go check your microwave or TV right now.)

Another way to calculate usage is to consult a table that lists typical wattages of household appliances and electronic devices. Figure 8-12a shows a typical appliance wattage chart. It gives the electrical consumption for each device.

Making Sense of Generators

Off-grid systems typically require a generator for backup power, but generators are also needed for routine battery maintenance, notably periodic equalization. How often homeowners need to run a generator depends on many factors, among them the size of the PV system, household energy use, and the amount of available sunshine. Gas generators are fairly inexpensive, but they burn costly fossil fuels (gasoline). They're also prone to break down, and many models are quite noisy. In addition, some units may not last more than 500 hours. Because of these factors, I recommend that homeowners oversize their solar electric systems — that is, that they install a larger PV array and a larger battery bank than they might ordinarily consider. This way, they won't need to supply backup power so often. A larger battery bank means that batteries aren't deeply discharged as often, which reduces the need for periodic equalization. Or, you may want to consider installing a wind system to make up for shortfalls that occur during cloudy periods.

Multiplying wattage of each appliance by the estimated time it is used each day gives you the daily electrical consumption in watt-hours.

From this analysis, you should be able to calculate your total daily energy requirement in watt-hours. Note, however, that some appliances or electric devices may be used more during some parts of the year than others. Lights, for example, are on more often during the winter than the summer because the days are shorter in the winter. The television may run less frequently during the summer because the kids are outside playing. Try to take such differences into account when computing energy consumption (Figure 8-12b).

Once you know how much electricity you consume, you need to perform some additional calculations. These are typically in a worksheet form. You can also complete this process using computer software.

The worksheets available through Solar Energy International's book *Photovoltaics: Design and Installation Manual* help you determine electric load (demand for electricity) as just described. You are then led through a few computations to determine the size of the inverter you'll need. Once you know this, you can determine the size of the battery bank you'll need if you are designing an off-grid home. The worksheets then lead you through calculations

to size the array — that is, to determine the number of PV modules you'll need. This calculation takes into account your demand, the efficiency of the entire system, and solar availability.

If your eyes are starting to glaze over, don't despair. A local installer can run the numbers for you. Even if you are planning on installing the system yourself, consider hiring a local installer to consult with you at this stage. (Be careful, if you install a system yourself. You'll be dealing with high-voltage DC and AC electricity that can be dangerous!)

Off-grid PV systems should be sized for the month with the highest energy demand. You can learn how to do this by taking a solar electricity course or by reading more advanced books on solar electricity. An installer will do the math for you, at no charge.

Grid-connected systems are sized for annual energy production. That is, you determine how much electricity you need annually. Once this is determined, you can size a system based on average daily use and the amount of sunshine in your region. Again, if your system is going to be installed by a professional, they'll run the math.

In an existing home, it is easy to determine the amount of electricity you consume each year. But in a new home, you'll need to make an educated guess.

Table 8-12a
Typical Wattage Requirements for Common Appliances

General household
Air conditioner (1 ton) 1500
Alarm/Security system.............3
Blow dryer.......................1000
Ceiling fan 10-50
Central vacuum 750
Clock radio 5
Clothes washer.................1450
Dryer (gas) 300
Electric blanket 200
Electric clock.......................4
Furnace fan 500
Garage door opener 350
Heater (portable) 1500
Iron (electric) 1500
Radio/phone transmit 40-150
Sewing machine 100
Table fan...........................10-25
Waterpik.............................100

Refrigeration
Refrigerator/freezer.............540
  22 ft³ (14 hrs/day)
Refrigerator/freezer.............475
  16 ft³ (13 hrs/day)
Sun Frost refrigerator..........112
  16 ft³ (7 hrs/day)
Vestfrost refrigerator/ 60
  freezer 10.5 ft³
Standard freezer 440
  14 ft³ (15 hrs/day)
Sun Frost freezer................112
  19 ft³ (10 hrs/day)

Kitchen appliances
Blender...............................350
Can opener (electric)...........100
Coffee grinder 100
Coffee pot (electric) 1200
Dishwasher......................1500
Exhaust fans (3) 144
Food dehydrator................600
Food processor...................400
Microwave (.5 ft³)..............750
Microwave 1400
  (.8 to 1.5 ft³)
Mixer..................................120
Popcorn popper.................250
Range (large burner) 2100
Range (small burner).........1250
Trash compactor..............1500
Waffle iron 1200

Lighting
Incandescent (100 watt) 100
Incandescent light 60
  (60 watt)
Compact fluorescent.............16
  (60 watt equivalent)
Incandescent (40 watt) 40
Compact fluorescent.............11
  (40 watt equivalent)

Water Pumping
AC Jet pump (¼ hp)............500
  165 gal per day,
  20 ft. well

DC pump for house 60
  pressure system (1-2 hrs/day)
DC submersible pump...........50
  (6 hours/day)

Entertainment
CB radio 10
CD player..............................35
Cellular telephone................24
Computer printer...................100
Computer (desktop).......80-150
Computer (laptop) 20-50
Electric player piano.............30
Radio telephone 10
Satellite system 45
  (12 ft dish)
Stereo (avg. volume) 15
TV (12-inch black & white)....15
TV (19-inch color) 60
TV (25-inch color) 130
VCR 40

Tools
Band saw (14")................1100
Chain saw (12")...............1100
Circular saw (7¼")..............900
Disc sander (9") 1200
Drill (¼").............................250
Drill (½").............................750
Drill (1").............................1000
Electric mower.................1500
Hedge trimmer 450
Weed eater........................500

Fig. 8-12: *Sizing a Solar Electric System. (a) This table provides typical wattage (power consumption) readings for major appliances and household electronics.*

(b) This worksheet can be used to list all of your appliances and to determine how much power they use each day.

But remember: reduce your demand first! Apply energy conservation measures vigorously before you estimate how much electricity you will be using.

BUYING A SOLAR ELECTRIC SYSTEM

When it comes to buying a solar system, many people wisely turn to a local PV

Table 8-12b **Stand-Alone Electric Load Worksheet**											
Individual Loads	Qty	X Volts	X Amps	= Watts AC	DC	X Use Hrs/day	X Use days/wk	+ 7 = days	Watts Hours AC	DC	
								7			
								7			
								7			
								7			
								7			
								7			
								7			
								7			
								7			
								7			
								7			
								7			
								7			
								7			
								7			
								7			
								7			

AC Total Connected Watts: _____   AC Average Daily Load: _____

DC Total Connected Watts: _____   DC Average Daily Load: _____

dealer/installer who can select the components and ensure that all of them work well together. Although this option may cost a bit more' than those I'll explain shortly, it's a good approach. A competent local installer can answer all of your questions and take care of problems that may arise. (Be sure they really know what they're doing.) Unless you're mechanically inclined and pretty knowledgeable, embarking on this process yourself can put you on a steep and treacherous uphill climb. "Even licensed electricians often need help from experienced PV installers," notes Weiss. You can find a list of local installers at homepower.com or at findsolar.com. Also, be sure to check out the local business directory your phone company provides.

Another approach is to buy a system from an Internet supplier. This approach can save you a substantial amount of money, as deep discounts are available through the Internet. Once the system arrives, though, you will have to install it yourself, or try to hire a local installer to do it for you. However, local installers may not be happy that you cut them out of the first part of the deal. (They make a little profit on the sale of equipment.) If you do buy through an Internet supplier, you'll need to know quite a bit more about PV modules, racks, inverters, charge controllers, disconnects, and batteries than when purchasing a system from a local supplier/installer. If you decide to take this route, you should read one of the books on solar electricity in the Resource Guide to deepen your knowledge. I've provided a lot of information in this chapter, but there's much more to know. As Weiss points out, "PV systems are not plug and play."

For years, my favorite book on solar electric systems has been *The New Solar Electric Home* by Joel Davidson. It presents a lot of technical information in a way that is amazingly comprehensible, and it was recently updated. Unfortunately, most of the books on solar electricity are penned by engineers or tech-heads for whom writing is not their strength. There is at least one exception, though, that might be useful to you: Johnny Weiss and his colleagues at Solar Energy International in Carbondale, Colorado have written *Photovoltaics: Design and Installation Manual*. This book is a manual for individuals who want to size, design, and install solar electric systems. It is generally well-written and full of good information. It should bring you up to speed on the subject so that you can size and design your own system and purchase components with confidence. I'd also strongly suggest taking a workshop or two to hone your skills. Hands-on experience is vital. You can take classes at The Evergreen Institute's Center for

Renewable Energy and Green Building, and a number of other places, including Solar Energy International in Colorado, the Solar Living Institute in California, or the Midwest Renewable Energy Association in Wisconsin. Many other educational programs have come on line in the past few years. Be sure instructors have experience and can explain this difficult subject clearly.

Be very careful when shopping for and purchasing the components of a solar electric system. This requires a lot of knowledge and attention to detail; be certain that you are working with a very knowledgeable dealer who really knows what he or she is talking about and offers solid technical support. Look for a supplier who's been around for a long time and who will sell you what you need, not what they have in surplus. In recent years, there's been an onslaught of Internet renewable energy suppliers. Some of them operate from remote sites; they have no inventory, so everything must be drop-shipped from the manufacturers. They typically offer little, if any, technical support. They lack expertise and may have lousy return policies. They may even charge you to replace items that they shouldn't have sold you in the first place. As a starter, I recommend that you visit Solatron Technologies' website (solaronsale.com) and read "Six Important Questions to Ask before Choosing an Alternative Energy Dealer." (As a side note, I've purchased PV modules, Water Miser battery caps, batteries, and an inverter through this company — and even sold a used inverter through their website — and have found their service and products to be exceptional. Moreover, their website is one of the most secure sites from which you can order. I also order a lot of components from Northern Arizona Wind and Solar. There. End of free advertisements.) Another highly reputable supplier is Real Goods. They've got excellent, highly knowledgeable people who can help you size and select components. Other online suppliers provide top-notch service and products, too, but you need to be careful when shopping for an online supplier. Look through *Home Power* and *Solar Today* magazines. Check out each company's website, its return policies, its expertise and level of technical support, and other aspects of their business. Does the company actually have a location, or are they operating from someone's home? Ask friends who may have dealt with online dealers for recommendations.

Buying Solar Modules

When buying PV modules, you'll find that you have quite a lot of options; numerous manufacturers produce several different sizes of modules. Fortunately, they're all pretty good, so it is hard to go wrong.

Given this, what criteria do you use to select a PV module?

If you are working with a local installer, he or she may have a favorite company they like to work with. In fact, most local installers are distributors for one or two of the major PV manufacturers. Call several local suppliers to see what each one offers. If you want more options and want to save some money, you can contact a reputable online source. Experienced and knowledgeable salespeople may be able to give you sound advice on which panels will work best for you. Be sure to ask for sale items, too. You can often obtain some amazing deals this way. Also, be sure to ask if their companies offer any solar kits. Solar kits are complete packages that include the modules, racks, inverters, controllers, and everything else you'll need. Kits usually include whatever type of module the supplier can order in bulk at a deep discount, which they pass on to you.

One criterion many solar buyers use when comparing modules from different manufacturers is the cost per watt. To determine this, simply divide the cost of a module by its rating in watts. The output in watts is determined under standard test conditions (1,000 watts per square meter of irradiance at 25°C/77°F cell temperature). By dividing the wattage by the cost, you can determine the cost per watt. While you're shopping around, be sure to check out the manufacturers' warranties. Most PV modules come with 20 to 25 year warranties.

A Short History of PV Cells

Although PVs are generally thought of as a modern invention, they were discovered a long time ago. In fact, they've been around for more than 100 years, according to John Perlin, author of several excellent books on the history of solar energy, including From *Space to Earth: The Story of Solar Electricity*. The very first cell was made by an American inventor, E. E. Fritts, from selenium with a transparent gold foil layer on top and a metal backing. When struck by light, this primitive cell produced a tiny electrical current. Fritts called his device a "solar battery." This mysterious invention stirred considerable controversy, though. Many prominent engineers and scientists at the time proclaimed that it violated the laws of physics. How could it make energy without burning a fuel? Although Fritts is credited with developing the first solar cell, it turns out that a Frenchman, the experimental physicist Edmond Becquerel, had made one in 1839 out of two brass plates immersed in a liquid. When exposed to sunlight, this unusual contraption produced an electrical current.

Today, most commercial PV modules are made from silicon, though new and more efficient designs made from materials with space-age sounding names are in the works.

Types of PV Modules

Solar electric cells come in three basic varieties, all made from silicon.

The first commercial solar cell produced was the single crystal cell. These were made from a pure ingot of silicon — a long, solid cylinder that was then sliced into thin wafers.

Unfortunately, producing pure silicon ingots requires lots of energy, and cutting the ingot into thin wafers produces lots of waste material that must be remelted and formed into new ingots. The process is expensive because it requires so much energy; however, single crystal solar cells (also referred to as "monocrystalline") boast the highest efficiency of the three solar cell technologies on the market today. Single crystal cells are about 15 percent efficient — that is, they convert about 15 percent of the sunlight energy striking them into electricity. Single crystal cells are still manufactured today, and are commonly used in PV modules by a number of companies, including Sharp. All the PV modules at The Evergreen Institute are monocrystalline.

Another, newer PV cell is the polycrystalline cell (Figure 8-13a). Polycrystalline cells are so named because they consist of numerous silicon crystals of varying size.

Fig. 8-13:
Polycrystalline Solar Electric cells:
(a) in production, (b) finished cell, and (c) modules in solar array.

BP SOLAR

BP SOLAR

BP SOLAR

They're beautiful to behold, but slightly less efficient than single crystal models — about 12 percent, compared to 15 percent. However, because they require less energy to manufacture, they may be a bit cheaper. They are used in many solar modules today.

Silicon solar cells can also be produced by depositing thin layers of silicon on a thin metal or glass backing. This technology is known as amorphous silicon. It uses less energy, and less silicon, but is less efficient than either the single crystal or polycrystalline cells — and they're only about 8 percent efficient.

Amorphous silicon technology was first used to create tiny solar cells for calculators and watches. However, amorphous silicon is damaged by direct sunlight and offered very low conversion efficiencies — only about 5 percent. In the years that followed the introduction of amorphous silicon solar cells, however, manufacturers have found ways to layer thin film materials to boost efficiency to about 8 percent and to make this material resistant to photodegradation. Today, thanks to this research, amorphous silicon is being used to produce solar modules. UniSolar produces amorphous silicon in long rolls with sticky backing that can be applied to metal roofing (Figure 8-14a). This is called PV laminate or PVL. It can only be applied to standing seam metal roofing. UniSolar also once manufactured solar cells that were incorporated in roof shingles, as in 8-14b, although this product has been yanked off the market — possibly for

Fig. 8-14: *Solar Roofing. Amorphous silicon ribbons can be used to make (a) long rolls of material that can be applied to standing seam metal roofing and (b) solar shingles. Both are forms of building-integrated photovoltaics.*

good — for a number of reasons beyond the scope of this book.

Thin coats of amorphous silicon can also be sprayed on glass, creating solar modules and solar electric window glass that produces electricity from sunlight. This application is ideal for skylights and glass canopies, like those over gas pumps at gas stations. It is being used in large commercial buildings, too.

Thin film glass and solar roofing materials are called building-integrated photovoltaics and are very popular these days. Be careful, though, because they do have some downsides. Solarized glass located on the south side of a building receives much less solar energy in the summer than the winter due to the sun's high angle in the sky. This can dramatically reduce a system's annual output. (Remember Dan's rule of technology: Just because something seems like a good idea, doesn't mean it is!) Although you may want to consider solar roof materials like PVL, if you are thinking about installing solar electricity and re-roofing your home, a coated glass skylight is probably not going to be an affordable option just yet.

Some manufacturers also sell solar roof shingles. These are small solar modules that take the place of roof tiles. They often blend very nicely with the roof, and are therefore popular among individuals who are interested in maintaining the appearance of their homes. They do have limited application, however, as most homes have conventional shingle roofs.

Buying an Inverter

Inverters are a key component of virtually all residential solar electric systems. They come in many shapes, sizes, and prices. When purchasing an inverter from a local supplier who's also going to install the system, your choices may be relatively limited. Installers typically have inverters they like — and trust — and will probably choose among those that fit your needs. Even though a local supplier/installer will do the thinking for you, you should still understand something about inverters so you don't get stuck with a model that doesn't work for you.

First and foremost, you need to determine whether you need an inverter for a grid-connected or off-grid system. Second, when selecting an inverter for a battery-based system, be sure that its voltage corresponds to the voltage of the battery bank. As noted earlier, battery banks in off-grid solar electric systems are wired at either 12-, 24-, or 48-volts (the most common are 24- and 48-volt systems; 12-volt systems are common in small applications such as cabins or summer cottages). In olden days, the array produced 12-, 24-, or 48-volt electricity, and the inverter would boosts the voltage to 120 volt AC electricity. In off-grid systems, the array, charge controller, battery bank, and inverter had to be wired to the same voltage. Today, thanks to advances in charge controllers, it is no longer necessary for all the components to match. You can, in fact, wire the array to a higher voltage to increase efficiency. The charge controller then converts that to DC electricity that is compatible — voltage-wise — with the inverter and battery bank. In these systems, the charge controller, inverter, and battery bank must all have the same voltage rating.

The next selection criterion is the wave output form. There are two types: *modified sine wave* and *sine wave*. Basically, the form of the output wave tells you how pure the electricity is. Sine wave is purer than modified sine wave (technically it should be called "modified square wave"). Sine wave is equivalent to the electricity you buy from the electrical grid. Sine wave inverters are also more expensive.

All grid-connected inverters must produce sine wave electricity. Off-grid inverters can be sine wave or modified sine wave. Unless money is a problem, I strongly recommend that you purchase a sine wave inverter, not a modified sine wave inverter. Sine wave inverters produce "cleaner" AC electricity, so they tend to work much better with modern electronic equipment. In fact, some of the newest

electronic equipment — like most of the energy- and water-efficient frontloading washing machines — won't operate on modified sine wave electricity. The sensitive computers that run these washing machine just plain won't work! Some laser printers also have a problem with modified sine wave electricity, as do some cheaper battery tool chargers. Furthermore — and here's an important thing — electronic equipment like TVs and stereos give off a rather annoying high-pitched buzz when they're operating on modified sine wave electricity.

When selecting an inverter for a grid-connected system, you will need to match the inverter to the array. More specifically, you need to be sure that the output of the array matches the input of the inverter (in volts). For off-grid inverters, you will need to check continuous output, surge rating, and efficiency.

Continuous output power is a measure that tells you how many watts an inverter unit can produce on a continuous basis (provided there's electricity stored in the battery bank). The Xantrex RS3000 inverter, for instance, produces 3,000 watts of continuous power, which means it can power a microwave using 1,200 watts, an electric hair dryer using 1,200 watts, and many other smaller loads simultaneously.

Surge rating is the wattage a battery-based inverter can put out over a short period, usually around five seconds. The Xantrex RS3000 inverter has a 7,500-watt surge power rating (60 amps). That means it can produce a surge of power up to 7,500 watts. This is important because many appliances and power tools (washing machines, refrigerators, and table saws, for example) require a surge of power when first turned on. It's required to get the ball rolling, so to speak — to start parts moving, overcoming inertia. Typically, these devices only need this power surge for a tiny fraction of a second, but without it, the tool won't start. (Kind of like people who need their morning coffee to get going.)

Next is efficiency.

Inverters consume energy to change DC to AC and to boost voltage. Efficiency should be high on your list whether you are buying a grid-connected inverter or an off-grid inverter. So, look for models with the highest efficiency possible. The Xantrex RS3000 has a peak efficiency of 90 percent. The Sunny Boy 2500U is 93 to 94.4 percent efficient.

If you are going to have a battery bank, you'll need an inverter that contains a battery charger. Today, all battery-based sine wave and modified sine wave inverters have the additional circuitry needed to charge batteries from an external AC source (a generator or utility power, for example). That's why they are called inverter/*chargers*. Even if you are not planning on installing a battery bank, you may want to consider purchasing an inverter with a battery charger — in case you later decide to add a battery bank for backup or go off-grid.

You should also check into noise, especially if the inverter is going to be installed inside your home. Be sure to ask about this feature upfront, and, if possible, ask to see a model you are considering in operation to be sure it's quiet. My Trace inverter, now manufactured by Schneider Electric, is located inside my home. The manufacturer describes it as quiet, but the unit emits a loud and annoying buzz. (If that's their notion of quiet, they must be deaf.) The first six months after I moved in, it drove me nuts, though now I'm used to it. The Sunny Boy inverters define the word quiet. Outback inverters and Fronius inverters I've installed are also extremely quiet.

When considering an inverter, there are a number of other things to look for: ease of programming, the type of cooling system in the inverter, and (for battery-based inverters) search mode power consumption. My first Trace inverter (DR2424, modified sine wave) was a dream when it came to programming: all of the controls were manual. I simply adjusted a dial to change the settings and I was done. My new Trace inverter (PS2524, sine wave) works wonderfully in all respects except for programming, which is a nightmare. I have found the digital programming to be very difficult, and the instructions don't help. You pretty much have to be a genius to figure them out. So find out in advance how easy it is to change settings, and don't rely on the biased view of a salesperson or a knowledgeable tech person. Ask friends or dealers/installers for their opinions or, better yet, have them show you.

Last, but certainly not least, is the power consumption under search mode. Search mode is an operation that lets an inverter operating off batteries to turn off completely when there are no active loads in a home — no devices or appliances drawing power. To stay on the alert, though, inverters send out tiny pulses of electricity that search for a load (a closed circuit). If you switch on an appliance, the PV system immediately kicks into gear and starts supplying electricity via the inverter to meet the demand. The search mode saves energy because it allows the inverter to shut down and go to sleep (with one eye open). This is especially beneficial at night — when you go to bed, or shut down your business.

A search mode consumption of under 0.5 amps is good. Be sure when buying an inverter that the supplier, be it an Internet supplier or a local vendor, takes the time to determine which inverter is the correct choice for you. Ask lots of questions.

Consult with experts. Or better yet, hire an experienced professional to install your system.

Supplying 240-Volt Electricity

Many homes require 240-volt electricity to operate appliances such as electric clothes dryers and electric stoves. As a general rule, you should try to avoid such appliances, not because a solar system cannot be designed to take care of them, but because they use lots of electricity, so you'll need many PV modules to supply them.

If you must have 240-volt AC electricity, don't despair. All grid-connected inverters produce 120- and 240-volt AC. Some battery-based inverters, like the Xantrex XW series inverters, produce 120- and 240-volt AC. For other inverters, you will need to wire two 120-volt AC inverters in series to produce 240-volt AC. You can also install a step-up transformer such as the Xantrex/Trace T-240. This unit takes 120-volt AC electricity from an inverter and steps it up to 240-volt AC.

Buying Batteries

If you are going to install a grid-connected solar electric system with a battery bank or an off-grid system, you'll need batteries. The more energy your home consumes and the longer the cloudy spells, the more batteries you'll need. (I'll say it again: be sure to cut your electrical demand through efficiency and other measures *first*. Efficiency will save you a fortune on PV modules and batteries.)

Most batteries used in solar electric systems are 6-volt, deep-cycle flooded lead acid batteries. Trojan L-16s have been the mainstay of the solar electric industry for years, but their dominance in the battery market has been challenged in recent years by several batteries from other companies, among them US Batteries and Rolls Batteries (Figure 8-15).

Batteries are wired in a combination of series and parallel circuits to produce 12-, 24-, and 48- volt systems. The 48-volt systems are the most efficient. As you will soon see, batteries are rated by their capacity to store electricity. The common measure for battery storage is amp-hours: an amp-hour is one amp of current flowing for one hour. A battery in a solar electric system should probably store 350 to 420 amp-hours of electricity to be useful.

When shopping for batteries, it's hard to go wrong. Most deep-cycle flooded lead acid batteries manufactured for solar systems are pretty good. But be sure to check out the storage capacity and manufacturer warranties first. The Surrette S460 comes with a seven-year warranty. The manufacturer will replace the battery free of charge for the first two years if it fails. After that, the manufacturer will replace it at a prorated value. Although lead acid batteries are less efficient than some of the newer battery technologies on the market today, old batteries are recycled. In fact, nearly 100 percent of the lead from used batteries makes its way back into the production cycle.

Solar electric systems can run on ordinary car batteries, but not for long. Car batteries are not designed for deep discharging — drawing off lots of power. They're the rabbits of the battery world; they're designed to crank out tons of amps to *start* a car. What you need is a tortoise — a battery that can give you all it has for long periods of time. No sprinters need

8-15: *Lead Acid Batteries. The Surrette battery shown here is an excellent choice for a solar electric system.*

SURRETTE BATTERY COMPANY LTD.

apply. So be sure not to make the mistake of running a solar electric system on car batteries. It is a waste of your time and money.

Also, be sure not to purchase marine deep-cycle batteries. They are only slightly better than car batteries. You'd be lucky to get more than a year or two of service out of this type of battery in an off-grid solar system. Like car batteries, marine batteries are manufactured to optimize cranking power, that is, they are manufactured with thin plates to provide a surge of power to start engines. Thin plates, however, are damaged by the deep discharges that typically occur in solar electric systems.

Golf cart and forklift batteries are better choices, because they contain many thick lead plates capable of undergoing deep discharges day after day. (But you have to remember to recharge as soon as possible after deep discharging!) A properly maintained golf cart battery could last three to five years in an off-grid solar electric system. If you can get them and get them cheaply, they might be a viable option. Bear in mind, however, that the typical golf cart battery only stores about two thirds of the electricity of a standard deep-cycle battery designed for solar electric systems.

Far better are the batteries manufactured specifically for renewable energy (RE) systems by companies like US

Battery, Rolls, and Trojan. These batteries have thicker plates that can withstand deeper and more frequent deep discharges than the standard golf cart battery. As a result, they have longer life expectancies — they'll last 7 to 15 years.

How long an RE system battery can last depends on how often it is deep discharged and how well you take care of it. Generally, the more often a battery is deep discharged, the shorter its lifespan. (They typically withstand 750 to 1,000 deep discharges before needing replacement.) If battery acid levels are not maintained by periodically adding distilled water, or if the batteries are not periodically equalized, expect a much shorter lifespan. Also, if batteries are kept in a cold place, they won't function optimally. They lose capacity when cold. Battery terminals need to be cleaned at least once a year, too. If they're not kept free of corrosion, batteries won't perform well. Better yet, coat terminals with a protective spray. I use a corrosion-resisting product like Permatex's Battery Protector and Sealer, sold in the automotive department of hardware stores. While you are at it, you should also be certain to keep the batteries clean. Wipe them off regularly for optimal performance.

Lead acid batteries contain sulfuric acid (a 30 percent solution). Although they work well if well taken care of, lead acid batteries produce hydrogen gas when they're being charged, whether by a solar array, a backup generator, or grid power (in a grid-connected system with battery backup). Lead acid batteries also produce a corrosive acid mist. They also need to be installed in a well-ventilated area away from people and sources of combustion because they release hydrogen and oxygen.

Whatever you do, don't buy used batteries. As Richard Perez of *Home Power* magazine says, used batteries are probably abused. Most of the people he knows who have installed used batteries ended up being sorry they did. To learn more about battery maintenance, I strongly recommend that you check out my book, *Power from the Sun*.

Sizing Your Battery Bank

Most professional installers design off-grid solar electric systems with about three to five days' backup, known as *battery days*. That is to say, they size the battery bank so that it can supply your needs for three to five cloudy days. Generally, the sunnier the climate, the fewer batteries you will need. A three to five cloudy-day storage capacity is best for most climates. While creating a more substantial storage capacity may give homeowners greater peace of mind, few beginners realize that PV modules produce electricity in cloudy weather, although at a reduced rate. The system doesn't go dead the minute a few clouds block the sun. To determine how many batteries are needed to provide sufficient backup, you'll need to run through some fairly complicated calculations that require computing your household's average daily consumption of electricity in amp-hours. Suppliers should be able to help you determine your needs, especially if they have worked with you to size your solar electric system. They'll also help you determine how many amp-hours of electricity you'll need to store in your battery system to ensure that you can achieve the desired number of days of autonomy.

Other Types of Batteries

Other than 6-volt, deep-cycle lead acid batteries, nickel cadmium and nickel iron batteries can also be used. These batteries can be deep discharged many more times than lead acid batteries and therefore last a lot longer. Unfortunately, they don't store as much electricity as a standard flooded lead acid battery, and they cost a heck of lot more. They're also not widely available.

Another type of battery that is useful in certain applications is the sealed battery. (They're also called *captive electrolyte batteries* for reasons that will be clear shortly.) To learn more about them, read the sidebar "Sealed Battery — A Misnomer?" Two types of "sealed" batteries are currently available: absorbent glass mat (AGM) batteries and gel cell batteries. In AGM batteries, thin absorbent fiberglass mats are placed between the lead plates to immobilize the acid. The mat is a microporous meshwork that creates tiny pockets that capture hydrogen and oxygen gases given off by the battery during charging. The gases recombine in these pockets, forming water.

In gel batteries, the lead plates are separated by cavities, as they are in a flooded lead acid battery. However, the electrolyte (sulfuric acid) is in a gel state, not a liquid. The electrolyte is gelled by the addition of a small amount of silica gel, which turns the electrolyte into a material much like hardened Jell-O.

Sealed batteries are maintenance-free, which means they don't need to be filled with distilled water or equalized. This saves lots of time and hassle and makes them a good choice for very remote locations where routine maintenance is unlikely, or for homeowners who don't want to be bothered by battery maintenance.

Sealed batteries are spill-proof. The gel cell batteries won't leak even if the battery casing is broken (a rare occurrence). Because of their design, sealed batteries charge faster than standard lead acid batteries. Sealed batteries release no explosive or toxic gases. In addition, sealed batteries are much more tolerant of low temperatures than lead acid batteries. They can even tolerate occasional freezing, although this is not recommended. Sealed batteries are commonly used for storing electricity in solar electric and wind-generating systems on sailboats and RVs where the rocking motion would spill the sulfuric acid of flooded lead acid batteries, and where space is limited and batteries are

Sealed Battery — A Misnomer?

Sealed batteries are so labeled because they have no caps, and you never have to fill them or fuss with the electrolyte. They also release no explosive or toxic gases, like conventional lead acid batteries. Basically, a sealed battery is filled with electrolyte at the factory, fully charged, sealed, and then shipped. Because they are sealed, they're easy to handle and ship without fear of spillage. A typical lead acid battery, on the other hand, must be shipped dry and filled with electrolyte (30 percent sulfuric acid) on arrival. The truth be known, sealed batteries are not totally sealed. Each battery contains a pressure relief valve that blows if a battery is accidentally overcharged. The more correct term for sealed batteries is "valve-regulated lead acid batteries" (VRLA). The valve keeps the battery from exploding. Once the valve has blown, though, the battery is done for, shot, kaput, dead.

frequently crammed into out-of-the way locations. They also have a lower rate of self-discharge, which means they discharge more slowly than conventional lead acid batteries.

Sealed batteries can be used for grid-tied systems with battery backup. In such instances, the batteries are typically kept at a full state of charge (they're regularly recharged by the solar array and the electrical grid). Unfortunately, sealed gel batteries are much more expensive than flooded lead acid batteries. They also typically store less electricity and have a shorter lifespan than the more commonly used lead acid battery. Steven Strong, solar electricity expert and author of *The Solar Electric House,* says, "They should be considered for all photovoltaic applications, especially those where site access for regular periodic maintenance is impractical." Frankly, I wouldn't recommend them for most home systems.

LOCATING A RELIABLE CONTRACTOR

In many locations, you need a permit to install a solar electric system. It's a good idea to check out permit requirements *before* you order any equipment. A local supplier/installer should be able to help you with this or even obtain the permit for you. While you are at it, be sure to check out financial incentives, too, before you order your equipment. Incentives may come with some restrictions imposed by local utilities or by the state government that will affect the size of the system you can install, the type of equipment, and the installer. For example, utility rebates apply only to grid-connected systems.

To locate a reliable installer, it's always good to ask around. Talk to people who have had systems installed recently. Get their advice on companies listed in the phone book or ask for references for local suppliers/installers. Professional credentials are one indication of a PV dealer's knowledge and qualifications, so be sure to ask what courses they've taken, what certifications they've earned, and what licenses they've received. Installers can receive certificates in PV systems through educational centers like mine, The Evergreen Institute. Or, they may have received degrees or certificates from two-year technical colleges. Today, more and more installers are certified by the North American Board of Certified Energy Practitioners (NABCEP). But just because someone is NABCEP-certified doesn't mean they have much experience. NABCEP certification can be obtained in a number of different ways. To qualify, you only need to install or be a part of a handful of installations.

So, check for credentials, but also look for experience. Be sure to find out how long they've been in the business and how many systems like yours they've installed. Ask to see the systems and talk to the owners. Service is important, too. Be sure they will be there to take care of your system if you have any problems with it. Find out what services your supplier/installer will provide. Also, will your installer provide any performance guarantees? What are the warranties offered by the manufacturers of the components of your system? Finally, be sure to get a cost estimate up front, and ask that the dealer itemize all aspects so there are no surprises. "Consider having a dealer supply, install, and warrant your system. Go for the full-service approach," advises Johnny Weiss.

WHY INSTALL SOLAR ELECTRICITY?

Solar electric systems are a great source of clean, reliable electricity. Solar electricity can be stored in battery banks for use at night or on cloudy days. Or it can be "stored" on the electrical grid. Remember, even though initial costs may be high — from $5,000 to $40,000 or more, depending on the size and the type of system — the fuel (sunlight) is free, abundant, and clean, and it's not under the control of some powerful multinational corporation. Financial incentives discussed earlier in the chapter and good net metering laws help make this venture more cost-effective, too.

Solar electric systems require minimal maintenance, too, unless batteries are included in the mix. And solar electric systems are a relatively environmentally benign source of energy. In addition, they operate quietly, unless you need considerable backup from a generator. Solar electric systems are modular, which means that you can install them incrementally. You can start small and expand as your finances permit. Solar electricity, like other technologies discussed in this book, provides energy independence. No matter whether you live in a busy city or a sleepy town or a remote rural area, solar electricity can meet your needs, and help replace waning supplies of oil and natural gas.

CHAPTER 9

WIND POWER

MEETING YOUR NEEDS FOR ELECTRICITY

In 2004, I took a group of students in my Introduction to Sustainable Development course at Colorado College to a coal-fired power plant in Denver. After trudging through the hot, noisy, and dangerous facility, listening to the spiel on the virtues of coal, we hopped in our vans and drove north to Wyoming to visit a wind farm. The wind farm was perched on a plateau visible from the interstate. As we drove onto the property, the giant white wind turbines came into full view.

Fig. 9-1: *These contented bison graze around the base of the gigantic wind turbines at the Ponnequin Wind Farm in northern Colorado, demonstrating that wind energy can be coupled with other commercial activities, notably farming and ranching, providing dual incomes for landowners.*

MATT REUER

Bison grazed around the base of the towers, munching on grass (Figure 9-1). My students were awestruck when we got out of the vans and stood at the base of these massive, quiet turbines towering 260 feet above our heads.

As I looked around, I could see smiles on my students' faces. The contrast between the noisy, dirty coal-fired power plant fed by huge piles of coal shipped in from hundreds of miles away, and the hushed efficiency of the wind farm fed by an abundant, clean, and free renewable resource was just too much for them. "Wow," one student said. "This is amazing!" burst another.

I was pretty amazed, too. It was my first up-close encounter with gargantuan commercial wind generators. Watching the shadows of the turning blades was indescribable — "way cool," as one of my young students remarked. I don't know how I could improve on his assessment. What is more, the turbines were remarkably quiet. All they made was a swooshing sound as the blades turned in the cold wind.

This chapter is not about large commercial wind systems; it is about smaller systems, typically referred to as *small wind* systems. These are systems that generate electricity for individual homes, businesses, farms, and ranches. This chapter should help you decide if wind is an option for you and if you have sufficient resources to make it work. It will also help you determine which type of wind turbine and wind energy system is right for you, and it will help you understand the importance of tall towers.

IS WIND POWER IN YOUR FUTURE?

Wind isn't for everyone. In fact, if you live in a city or town, chances are wind power is not going to be an option for you. Although the wind blows in cities and suburbs — sometimes quite fiercely — in heavily populated, built-up environments, wind flows are often extremely turbulent. That is to say, although the wind blows, it doesn't flow smoothly. It's like water in a rapids.

Wind can't flow smoothly in built-up areas, because there are too many obstructions — trees and buildings — that generate turbulence. Turbulence is pure hell on wind turbines, viciously tearing many of the lighter-weight models apart. (Turbulence is to wind turbines as potholes are to cars.) Even if turbulence weren't a problem, there's often not enough room in urban and suburban neighborhoods to install even the smallest wind turbines. And even if there is room, chances are neighbors will complain, saying the turbine is unsightly. Or city officials might tell you that wind turbines violate height ordinances, which restrict structures to 35 feet.

Wind power is primarily useful as a source of electrical energy in rural areas on lots of one acre or more, preferably two or more, but then only in locations where there are no ordinances that prohibit installation of wind turbines. Even though that limits the potential for residential wind energy, there are still plenty of suitable sites. According to *Small Wind Electric Systems* published by the US DOE, "Twenty-one million homes in the United States are built on one-acre and larger sites." So, about 10 percent of the US population could theoretically take advantage of wind energy. But that number may be exaggerated. Wind power will work only for those who live in rural areas with sufficient wind resources. I've found that it is difficult to site wind turbines on small lots. You need at least two acres, preferably more, to find a suitable location — one that is far enough away from neighbors and roadways, buildings, and electric lines. If you live in a city or town, though, don't be disheartened. You can still power your home with wind energy by buying wind power from your local utility. I'll explain how at the end of this chapter.

WIND POWER: A BRIEF HISTORY

Unbeknownst to many, winds are generated by solar energy. For example, in coastal areas, offshore winds are created by sunlight warming the land masses. Hot air, created when sunlight strikes the land, rises. Cool air from nearby water bodies — large lakes or the ocean — rushes in to fill the gap, creating fairly reliable winds.

Even large air movements across entire continents are driven by the differential heating of the Earth's surface, which is cooler at the poles than at the equator. Hot air rises at the equator and cooler air from the poles moves in to replace it. The result is huge air circulation patterns — what meteorologists refer to as *prevailing winds*.

Humans have tapped the power of wind energy for centuries, first to power sailing vessels and later to grind grain and pump water. It is believed that the very first wind generators were used in Persia in the fifth century AD. Wind turbines spread from the Middle East to Europe in the 12th century AD. From Europe, wind technology spread to North America; by the late 1800s, there were thousands of windmills in Europe and North America in rural settings. From the late 1880s to the early 1900s, more than eight million windmills, most of which were used to pump water for livestock, were installed in the United States, according to the National Renewable Energy Laboratory (Figure 9-2). Some rural farmers installed electric-generating wind machines, known as *wind turbines*, to provide electricity for lights, toasters, and radios. In the 1920s, however,

American cities began to electrify thanks to the pioneering work of Nikola Tesla, the brilliant scientist who invented equipment that made AC electricity possible.

DAN CHIRAS

Fig. 9-2: *This old windmill, which is still functional today, was once a common sight on the plains of North America, where it was used to pump water for livestock. Its larger cousin, which produces electricity in the background, is now becoming a much more common sight throughout the world.*

At first, the electricity that powered cities came from centralized power plants that burned coal. Electric lines strung here and there like spider webs throughout cities delivered electricity to homes, offices, and factories. By the end of the 1920s, North American cities were completely electrified, and oil lamps were a thing of the past. But things were quite different in rural areas. There were no central power plants to deliver electricity to rural residents. And there were no electric lines to bring electricity from cities. Small wind turbines were the sole source of electricity on farms. The US government then set its sights on rural areas, launching the Rural Electrification Program. As a result of this ambitious program, wind power entered a period of steady decline. In fact, farmers were required to tear down their electric-generating wind turbines as a condition of connecting to the new electric grid. Wind energy did not die out completely, though. Many small windmills continued to pump water to fill stock tanks on the Great Plains, and some are even working today. (Some areas of the Midwest weren't electrified until the 1970s.)

Then came the 1970s and oil shortages. Although the crises were artificial (the result of the actions of the oil-producing countries rather than actual depletion of Earth's resources), they fueled tremendous interest in energy self-sufficiency, conservation, and renewable energy. Wind benefited greatly, even though it was largely liquid fuel supplies that were threatened by the oil embargoes of the 1970s.

By the early 1980s, there were 80 companies in the United States manufacturing windmills and wind turbines, including large commercial generators and smaller household-sized units. During this period, several utility companies in California built large wind farms to provide electricity to residents in San Francisco. Unfortunately, early wind turbines were plagued with mechanical problems and frequent breakdowns. Large commercial wind turbines, for example, could only be counted on 60 percent of the time. They were broken down and out of service the rest of the time. The early models were also fairly inefficient, too. These problems, combined with a growing lack of concern over achieving energy independence, caused wind power to decline once again. Moreover, because of wind energy's bad reputation, many people lost interest in this important technology, and some people today continue to view wind power with undeserved skepticism. However, convinced of wind's importance, several manufacturers tackled the technological problems that arose in the 1970s and early 1980s and have greatly improved their products.

Thanks to the efforts of these manufacturers, often assisted by the National Renewable Energy Lab in Golden, Colorado, the efficiency of large commercial wind turbines has increased, climbing two or three times above efficiencies achieved in the 1970s and early 1980s. Reliability has increased as well, rising from 60 percent to an astounding 97 to 99 percent. Small wind generator manufacturers like Bergey (based in Oklahoma) improved the design of their turbines, simplifying them and getting rid of parts, like the brushes, that required frequent replacement. These improvements have created a resurgence in wind power production. Today, wind power is the fastest-growing source of energy in the world.

UNDERSTANDING WIND GENERATORS

Wind generators go by several names: wind turbines, wind machines, and wind plants. The term *windmills* is generally reserved for water-pumping wind machines.

Residential wind generators, both large and small, have the same three basic components: (1) a blade assembly that turns in the wind, commonly referred to as a rotor; (2) a shaft that connects to the rotor and rotates when the blades turn; and (3) a generator, a device that produces electricity (Figure 9-3). Many modern wind turbine manufacturers now directly connect the rotor to the generator, eliminating the shaft.

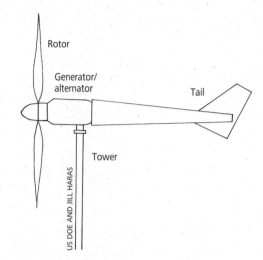

Rotor

Generator/
alternator

Tail

Tower

US DOE AND JILL HARAS

Fig. 9-3:
*Anatomy
of a wind
turbine.*

Wind turbines exist in two basic varieties: horizontal axis units and vertical axis units, as shown in Figure 9-4. Horizontal axis units are the most widely used in household-sized wind systems. Vertical axis wind turbines, which are wildly popular among newcomers and new inventors, have blades that resemble big egg beaters. When wind strikes the blades, they rotate around a vertical shaft. Cool as they are — and there are many very neat designs — these turbines are typically mounted on short towers, where there's very little wind.

AFRICAN WIND POWER

SANDIA NATIONAL LABORATORIES

Fig. 9-4: *Two types of wind turbine are in use today: the horizontal axis turbine shown on the left, and the much less common vertical axis turbine on the right.*

They're also less efficient than horizontal axis wind turbines. As a result, they're not typically worth the investment.

Because horizontal axis wind turbines are the main players in home wind energy systems, we'll focus our attention on them. From this point on, when I use the term wind turbine, I'll be referring to horizontal axis turbines. The turbines we'll be studying are typically used to provide backup power or to power homes, small businesses, or farms and ranches. Those wind turbines that are designed for battery-based systems are referred to as *battery-charging turbines*. Those that are designed for grid connection are known as *batteryless grid-tied turbines*.

A horizontal axis wind turbine is equipped with two or three blades, three being the most commonly used and the most desirable: the use of three blades results in less wear and tear on the generator in shifting winds and results in a more durable, and hence reliable, wind turbine. Blades are typically made of highly durable plastic, fiberglass, or rarely wood with a urethane coating. Many blades come with vinyl or polyurethane tape to protect the leading edge of the blade from wear. The blades of a wind turbine and the central hub to which they are attached are called the rotor. This is the "collector" of the wind turbine. The rotors of wind turbines capture kinetic energy from the wind and convert it into rotating mechanical energy in the form of a spinning shaft. The spinning shaft is attached to the generator. It converts mechanical energy into electrical energy.

Several types of generators are found in household-sized wind turbines. The most common is known as a *permanent magnet alternator*. They are called permanent magnet alternators because they contain real magnets (as opposed to electromagnets). When the blades of a wind turbine spin, the magnets typically rotate around a number of tightly coiled copper wires, called *windings*. The movement of the magnetic field past the windings causes electricity to be produced inside the windings. (If you remember your high school physics, moving an electrical coil through a magnetic field creates an electrical current in the wires.)

Wind turbines produce "wild" AC electricity. This is alternating current electricity whose voltage and frequency (cycles per second) vary with the speed of the spinning blades (rpm) and thus the wind speed. Wires attached to the alternator transport the wild AC electricity to a device known as an inverter/controller. The controller converts the wild AC to DC. The inverter converts it back to 60 cycle per second 120-volt AC electricity. (You can learn more about this process in my book, *Power from the Wind*.)

WIND SYSTEMS:
THREE BASIC OPTIONS

Like solar electric systems, wind systems fall into three basic categories: (1) grid-connected, (2) grid-connected with battery storage, or (3) off-grid. As you may recall from Chapter 8, a grid-connected system consists of three main components: (1) a technology that captures some form of renewable energy, in this case, a wind turbine; (2) an inverter, and (3) a main service panel or breaker box. As just noted, AC electricity produced by the wind turbine is converted to DC by the controller inside the inverter, and then is converted to AC electricity. It flows to the main service panel. From here it travels to various active circuits to power devices that consume electricity, known as loads. As in a grid-connected PV system, excess power flows onto the grid, and may be "banked" as a credit on your utility bill. When the wind isn't blowing, the electricity you need to power your home comes from the grid. Because the grid is your "storage battery," a grid outage will result in a power outage in your home even if the wind is blowing. This is due to the design of the inverters. (For more details, read the section on grid-connected solar electric systems in Chapter 8.)

If you want to store electricity for emergency use — for example, to protect your home, office, or farm from occasional outages — you can install a battery bank for backup power. In these systems, the batteries are kept full all the time, in case there's an outage. If a blackout occurs, the inverter converts to battery operation, and DC electricity is drawn from the batteries. The inverter converts the DC into AC, thus supplying electricity to your home to keep the refrigerator running and lights burning. In most grid-connected systems with battery backup, the size of the battery bank is quite small — only enough to supply critical loads during power outages. (For more details, check out the section on grid-connected systems with battery backup in Chapter 8.)

Your third option is an off-grid system, that is, a system that provides 100 percent of your electricity. Off-grid systems are not connected to the electrical grid. Because of this, you cannot send surplus to the grid in times of excess production, nor can you draw from the grid in times of need. In off-grid systems, surplus electricity is stored in a bank of batteries, usually lead acid batteries. (This battery bank is much larger than in the previous system.) Electrical demand during windless periods is satisfied by electricity stored in the batteries, as in an off-grid PV system. A backup electrical generator may also be required. (Again, you can refer to Chapter 8 if you need to brush up on off-grid systems.)

Hybrid Systems

Winds tend to blow in the fall, winter, and spring, then die down over the summer. Because of this, homeowners and business owners who want to generate all of their electricity from renewable sources often install hybrid systems. A hybrid system is a renewable energy system that combines a wind turbine with solar electricity (PVs) or some other renewable energy technology. PVs and wind work particularly well together in many parts of the world because, in many locations, winds pick up in September or October, and continue through April or May. The strongest winds occur between November and March. Sunlight, on the other hand, tends to be strongest in the summer — the period when winds are weakest (Figure 9-5). A hybrid system consisting of a wind turbine and PV array can supply 100 percent of a family's, business's, or farm's electrical needs. During the summer, the PV system provides most of the electricity. The wind turbine is the backup source. During much of the rest of the year, the wind system produces most of the electricity, with the PV array providing supplemental power.

Although wind and solar electricity work well together and can provide virtually all of your electricity, you may still need to install a backup generator to supply electrical energy in periods of low wind

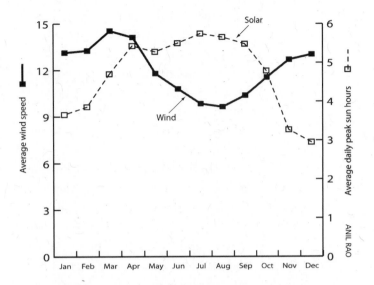

and low sunshine. (Or, you can simply cut back during these periods.) You also need a gen-set to maintain batteries in peak condition, a topic discussed in Chapter 8.

IS WIND ENERGY APPROPRIATE WHERE YOU LIVE?

Before you invest too much time learning about wind generators and wind energy systems, it is important to ask a key question: Do you have enough wind energy in your location to make an investment worth the money?

Assessing Your Wind Resource

Wind is the clean, free fuel that powers wind turbines. In order for a system to make sense economically, you need a site with sufficient wind. Most systems

Fig. 9-5:
Wind and solar resources are often complementary, as shown here and explained in the text.

for homes and farms require an average annual wind speed *at ground level* of 7 to 9 miles per hour. The higher, the better.

This does not mean that wind speeds need to blow at 7 to 9 miles per hour year round. It means they need to average 7 to 9 miles per hour at ground level. However, wind turbines are not mounted at ground level. They're typically mounted on sturdy towers 80 to 120 feet tall — sometimes even higher. At these heights, winds blows at higher speeds. For example, a wind blowing at 8 miles per hour at ground level could be blowing at about 13 or 14 miles per hour at 100 feet above ground. (Winds also blow more smoothly at this height because there is less turbulence.) Slight increases in wind speed dramatically increase the amount of power available from the wind. For example, increasing wind speed from just 8 to 10 miles per hour (a 25 percent increase) will increase the power available by 100 percent! The more powerful the wind, the more electricity a wind turbine can produce.

As a rule, the areas with the best wind resources in North America tend to be along seacoasts, on ridgelines, on the Great Plains, and along the Great Lakes. The northeastern United States and the deserts of the southwest are also excellent wind sites. That said, many other areas have sufficient wind resources to make a wind system a viable option. Before even

considering installing a wind system, you should assess your site very carefully — or better yet, hire a professional wind site assessor to do the job for you.

To assess the suitability of your location, you can begin by taking a look at a state wind map. You'll find your state's map on the Wind Powering America website (windpoweringamerica.gov/). The map you will find there shows average wind speeds, but often at 165 to 198 feet (50 to 60 meters) above the surface of the Earth. Wind turbines are typically a bit lower, so a wind site assessor will extrapolate downward using a mathematical equation to determine the average wind speed at the hub height of the turbine and tower you are thinking about installing. I present this information in my book, *Power from the Wind* and in my courses on wind site assessment. In Canada, you may want to check out the Canadian Wind Atlas at windatlas.ca/en/maps.php.

Don't make a decision based on a map, however, because local terrain may cause the wind resource at a specific site to differ considerably from estimates. Wind resources often vary significantly over an area of just a few miles because of hills and valleys, ridges and other factors. If you live in a valley, for instance, your average wind speed will probably be much lower than indicated by the map. If you live on top of a hill and aren't surrounded by tall trees, the average wind speed may be higher than the wind maps indicate. This is why you should hire a certified wind site assessor. Using some sophisticated tools and websites, he or she can estimate what the actual wind speed is.

When I do wind site assessments, I obtain my data from NASA's website, Surface Meteorology and Solar Energy. This site allows you to enter the height of the proposed tower and will give you an average expected wind speed for each month and for the year. Be sure to examine average wind speed by month, so you know how much wind is available during different parts of the year. This is especially important if you are planning on installing a wind/PV hybrid system.

If you or the wind site assessor determines your site is suitable for wind — that is, the average wind speed is sufficient — you need to determine if there is a location on your property where you can install a tower so that the turbine will fly high enough above all obstructions. The rule of thumb is that a wind turbine should be located so the entire rotor is at least 30 feet above the closest object within 500 feet. If the tallest object is a tree, take into account tree growth. As an example, suppose your property had a 50-foot barn, a 60-foot silo, and a 55-foot tree that matures at 75 feet. What should the hub height of the turbine be?

To determine hub height, start with the 30-foot rule. The blades of the rotor should be at least 30 feet above the tallest object within 500 feet to reach the smoothest laminar winds (winds that blow without turbulence). In this case, the silo is the tallest object, but the tree will grow to 75 feet. The rotor should therefore be located 105 feet above the mature height of the tree. If the blades on the turbine are 10 feet long, the hub height (the height of the top of the tower where the hub of the rotor is located) should be 115 feet. Since towers come in 20 foot increments, your tower will need to be 120 feet tall.

SELECTING A WIND GENERATOR AND TOWER

If wind is a viable option for you, you'll next need to determine which turbine and tower to buy. You've got lots of choices. This decision typically hinges on two basic considerations: how much electricity you want to generate from the wind and how much money you have to spend. As you shall soon see, there are hidden traps for those who don't shop carefully. I'll explain how you can avoid these traps in this section.

Choosing a Wind Generator

As just noted, choosing a wind generator can be tricky. There are numerous manufacturers and numerous models to choose from, ranging from 400-watt to 20,000-watt units (see sidebar, "Wind Turbine: Which Size is Right for Your Home?"). Moreover, there are many different factors to consider when making a decision, among them rotor size, cut-in wind speed, rated output, swept area, weight, and price. There's also the issue of sound and reliability. Finding information on the latter may be difficult. As you shall soon discover, not all of the information provided on wind turbines by manufacturers and retailers is useful. Bear in mind, though, if you are dealing with local wind energy suppliers/installers, they can often ease the pain a bit. A reliable supplier/installer, for example, typically recommends a few models with which he or she has had success. Be sure to hire someone whose opinions are based on experience — not what they've read in the product literature or been told by company sales staff.

Wind turbines can be purchased through national suppliers like Gaiam Real Goods or online suppliers. Although such sources typically carry models that work well for them, you should know as much as you can about wind turbines before putting your money down. This section will help you understand the most important factors you'll need to consider when shopping for a reliable wind turbine. If you want to do a detailed analysis of your options before talking to a local dealer/installer or before perusing catalogs of various online suppliers, I strongly recommend that you read my books *Wind Energy Basics* or *Power from the Wind*, or attend a few small wind energy classes.

Swept Area

As you may recall, the spinning blade assembly is known as the rotor. It is the

Wind Turbine: Which Size is Right for Your Home?

Wind generators for use in homes, on farms and ranches, and in small businesses, come in many sizes, ranging from 400 watts to 20,000 watts. This measurement is their output at a certain wind speed, known as their *rated wind speed,* and serves as a rough guide for selecting a turbine. Generally, the 400- to 1,000-watt turbines only supply 40 to 200 kWh of electricity per month, and only in areas with good, solid 12-mile-per-hour average wind speeds. Most homes in America consume around 750 to 1,000 kilowatt-hours per month. All-electric homes could easily consume 1,500 to 2,500 kWh per month, as would a small business or a farm or ranch operation.

Unless your home or business is super-efficient, you will need a larger wind turbine, around 2,500 to 6,000 watts. Even then, you'll need to use electricity efficiently unless you install one of the largest and costliest models with rated power values of 10,000 kWh. Or you could install a hybrid system: a smaller wind generator with a PV array.

collector of wind. That is, it converts the movement of air passing by the blades into mechanical energy, which is, in turn, converted into electrical energy by the alternator. As a general rule, the larger the rotor, the more electricity it will produce. As Mick Sagrillo points out in his piece in *Home Power*, a wind generator's rotor size "is a pretty good measure of how much electricity a wind generator can produce."Although other features such as the efficiency of the generator and the design of the blades can influence efficiency, Sagrillo argues that "they pale when compared to the overall influence of the size of the rotor." With wind turbines, size does matter.

Rotor size is determined by the length of the blades, which represent the radius of the circle described by the spinning blades. The area of that circle is known as the swept area. The swept area is measured in square feet or square meters. Figure 9-6 illustrates the differences in swept area of three household-sized wind turbines. You can quickly pick out the high-energy models just by looking at the swept area. This is the one factor that allows for easy comparison of different models.

Cut-In Speed

Household-sized wind turbines start producing electricity at wind speeds in the range of six to eight miles per hour. This is

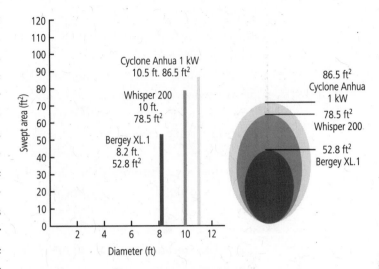

known as the *cut-in speed*. Unfortunately, they don't start producing useful amounts of electricity until wind speeds of 10 to 12 miles per hour. As a result, cut-in speed is pretty meaningless. Nevertheless, many people are impressed by turbines with low cut-in speeds. Don't be one of them. Although these turbines may start producing some electricity at low wind speeds, it won't be a significant amount.

Rated Output

As already discussed, PV modules are compared by their output — the amount of electricity they produce under standard conditions. Many modules these days produce 200 watts, give or take a few, under standard test conditions. Wind turbine manufacturers also rate the output of their turbines in watts. Bergey Windpower's

Fig. 9-6:
The swept area varies dramatically from one wind turbine to the next. By and large, you should choose the model with the greatest swept area. It will produce the most electricity at your site.

BWC XL.1, for instance, has a rated power output of 1,000 watts. So does Southwest Windpower's Whisper 200.

The trouble is, there are no set standards for determining rated output in the wind industry, as there are in the PV industry. (This could change soon, as efforts are underway to implement standards for rating wind turbines.) For example, the two manufacturers just mentioned rate their wind turbines at different wind speeds (called *rated wind speeds*). Bergey rates their turbine at 24.6 mph, while Southwest Windpower rates theirs at 26 mph. So which one is better? Because they are rated at different wind speeds, you can't really say. You would think that the wind turbine that produces 1,000 watts at the lowest wind speed would be the best buy. But not in this case. To make a decision, you'd be better off looking at the swept area. In this case, the swept area of the Bergey is 52.8 square feet and the Whisper is 78.5 square feet. Which one will produce the most electricity for you? The model with the largest swept area. As my friend and wind mentor Mick Sagrillo says: "While comparing PVs based on rated wattage makes for great cost comparisons, comparing rated outputs is a poor way to compare wind generators. You are far better off comparing swept areas or the kWh per month of electricity the different systems will produce at different average

wind speeds." In the example just given, the Bergey produces 115 kWh per month in an area with 10 mph wind, while the Whisper produces 125 kWh per month. The monthly energy output at various wind speeds is a much better criterion. I typically compare wind turbines on the basis of annual energy output, or AEO, at various wind speeds. This tells you which turbine will produce the most energy.

Governing System

All wind turbines come with a mechanism to prevent the generator from burning out in high winds. Why?

High winds increase the rpm of a wind turbine, which increases the amount of power produced. This is generally desirable. However, really high rpm can lead to overheating and burnout. The windings can get so hot they melt. To prevent this problem, manufacturers install some type of governing system also known as *overspeed control*.

One form of overspeed control simply turns the blades out of the wind so they slow down or completely stop spinning. The blades may tilt up and out of the wind. This is referred to as *top furling*, and is shown in Figure 9-7a. Or, the turbine may fold, so the blades turn out of the wind. This is referred to as *side furling*, and is shown in Figure 9-7b. Both mechanisms reduce the frontal area of the turbine, that is, the amount of rotor surface facing the wind. This, in turn, reduces the swept area, and slows the blades down to prevent the turbine from spinning too fast.

The other governing system changes the blade pitch, that is, the angle of the blades, so they no longer intercept the wind as efficiently. This causes the rotor to spin at a slower rate. Although blade pitch governing systems allow wind turbines to continue to produce power, they require more moving parts. As a rule, the more moving parts you have, the more maintenance. Talk to local installers for their recommendations.

Shut-Down Mechanism

Most wind turbine manufacturers include a shut-down mechanism — a mechanical or electrical device that allows an operator to shut a wind generator off. This is important because it allows the owner to repair or maintain the wind turbine without fear of injury. It also provides a means to shut the wind turbine down when a violent storm is approaching. Shut-down mechanisms come in many varieties, from disc and dynamic brakes to folding tails — tails that fold in such a way that the wind turbine is forced out of the wind and the blades stop rotating. The shut-down mechanism of a wind turbine is a key factor to consider when shopping for a wind turbine.

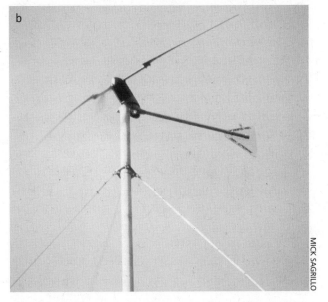

Fig. 9-7 a and b: *Overspeed control. (a) This wind turbine from North Wind Power top furls. Notice how the rotor rotates upward to reduce swept area and prevent the machine from spinning too fast. (b) This wind turbine side furls to achieve the same effect. Side furling is more common.*

Unfortunately very few turbines have *reliable* shut-down mechanisms. The most effective seem to be those that allow the operator to crank the tail out of the wind or engage a disc brake. This is achieved by turning a crank at the base of the tower.

Sound Levels

Wind turbines are not, by their very nature, quiet, although some models are significantly quieter than others. In fact, some of them are so quiet, you have to look to see if they are actually spinning. Sound is primarily generated by the blades cutting through the air and the spinning of the internal components of the generator.

As a general rule, the blades of lightweight wind generators spin faster than blades of heavier units. The higher the rpm of the blades at the rated wind speed, the more sound a turbine will make.

Fortunately for all concerned, the highest sound levels occur at high wind speeds, when background noise increases as well — for example, when the leaves and branches of trees are being blown by the wind. Also, wind blows by your ears, making it more difficult to hear. Also, fortunate is the fact that sound drops off fairly quickly with distance. Thus, the taller the tower, the less noise you will hear. Even so, I can still hear my 2.5 kW Skystream any time it is spinning, except in really strong winds which drown out the turbine, anywhere on my 50-acre educational center/farm — and it's on a 126-foot tower. So, be sure to listen to the turbine or turbines you are thinking about buying *before you buy* to assess sound levels. To minimize disturbance, install a turbine as far away as possible from your home and neighbor's homes.

The crank is attached to a cable that either engages the brake or causes the wind turbine to fold on itself, which pulls the blades out of the wind.

Durability

The durability of a wind turbine is important criterion when shopping. The more durable the machine, the longer it will last, the less maintenance it will require, and the more electricity it will produce. In short, a more durable turbine is a better investment of your time and money.

You might think that one metric published by wind turbine manufacturers — maximum design wind speed — would be a good indicator of durability. However, as Mick Sagrillo says, "it has little bearing on the expected life of a wind generator." Sagrillo points out that wind generators are designed to survive wind speeds of 120 mph or more, but they are not necessarily tested at these speeds — or repeatedly tested at these speeds — to see if the claims really hold up over time. And, in fact, more turbines are damaged by turbulence than high winds.

Sagrillo argues that the best criterion for durability is tower top weight: how much the unit weighs. "My experience," says Sagrillo, "is that heavy duty wind generators survive, and light duty turbines do not." Therefore, even though most wind turbines are rated for 120 mph or greater maximum wind speeds," experience indicates that many of the lighter turbines cannot handle sites with heavier winds or turbulence." Of the two models already mentioned, the Bergey XL.1 weighs 75 pounds and the Whisper 200 weighs 65 pounds. Neither is very heavy, but the Bergey has a reputation for being a very durable wind turbine. However, says Sagrillo, "Be forewarned! Weight … will be reflected in the price." As a rule, the heavier the unit, the more it will cost.

When you are shopping for wind turbines, you will often encounter power curves, like the one shown in Figure 9-8. A power curve is a graph that plots the power production in watts for a given wind turbine at different wind speeds. While they are nice to look at, they don't really help you very much in making a decision. Although 10,000 watts peak power on the curve looks impressive, remember that most turbines operate at the left end of the power curve — in slower, 10 to 15 mph winds. As you will see by looking at power curves, the output of a turbine at these wind speeds is usually pretty low. Again, the more valuable measures are swept area, annual energy (kWh) production per year at various average wind speeds, and weight.

Purchasing a Tower

A wind turbine constitutes a major expense in any wind system, but that cost is often rivaled by the cost of the tower. In this section, we'll examine towers.

Towers come in three basic varieties: free-standing, guyed, and tilt-up. Each type has a couple of subtypes.

Free-Standing Towers

Free-standing towers are those that require no additional support: they are self-supporting. Because they have a relatively small footprint, they are ideal for applications where space is limited. For example, if you only have an acre of land.

The most common free-standing tower is the lattice or truss tower, like the one shown in Figure 9-9. Lattice towers

Fig. 9-8:
This power curve shows the output of an ARE110 and an ARE442 wind turbine in watts (power) at various wind speeds. Note that wind turbines typically operate at the low end of the curve, in winds under 20 miles per hour.

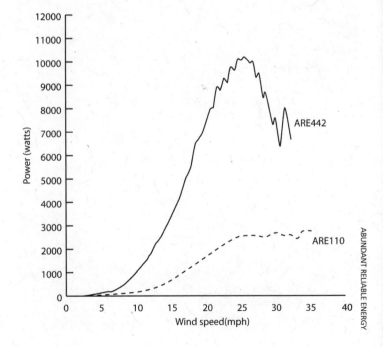

ABUNDANT RELIABLE ENERGY

Fig. 9-9:
The lattice or truss tower is strong and durable and an excellent choice for mounting larger wind turbines. It can be supported by guy wires, if necessary.

resemble the latticework of arbors. The Eiffel Tower in Paris is an example of a lattice structure. Lattice towers are secured to a massive foundation and are engineered to withstand powerful winds. Because of the large amount of metal that goes into them and because of the need for a sturdy anchoring foundation, they are one of the most expensive tower options.

Free-standing lattice towers are typically installed in 20-foot sections. The sections are usually preassembled and welded or bolted together prior to delivery. When the sections arrive, they are bolted together on the ground, then lifted onto the foundation via a crane, and bolted in place. Some installers assemble lattice towers in the upright position piece by piece using a device known as a *vertical gin pole*. This is risky and time-consuming and generally avoided by experienced installers. Yet another method is to install a lattice tower with a hinge at the base. The tower is assembled on the ground, the wind turbine is attached, and then the tower is tilted up using a crane, thanks to the hinged base.

Another option for a free-standing tower is a monopole. This tower consists of rigid metal steel pipe. It is assembled on the ground, 20 feet at a time, then lifted and bolted onto a hefty concrete foundation using a crane. Monopole towers require a lot of steel and a very large foundation; therefore, they are the most expensive type of tower.

Whatever choice you make, be sure that the tower is strong enough to withstand both the winds in your area and the weight of the wind turbine — both of these can be quite substantial. Several of the largest home-sized wind turbines weigh around 1,000 to 1,500 pounds. The Jacobs 31-20, admittedly the heavyweight of the residential- or business-sized units, weighs 2,500 pounds!

BERGEY WINDPOWER

The Mathematics of Wind Power

The power available to a wind turbine (and thus, its electrical output) depends on many factors: the most important are air density, swept area of the rotor, and wind velocity. The relationship among these factors is expressed by the mathematical equation $P = \frac{1}{2} d \times A \times V^3$. (In this equation, P stands for power in watts; d stands for air density. A is the swept area of the rotor, and V is wind speed.) What this equation tells us is that the greater the density of the air, the swept area, and the wind speed, the more power is available in the wind (measured in watts).

Air density varies with humidity levels and elevation above sea level. Although you can't control air density, it is important to understand the role it plays in wind power generation. For example, air is less dense at higher elevations, so if you live in the mountains, you should lower your estimates of your wind turbine's expected output. Don't expect it to produce as much electricity as it would at a lower elevation at any given wind speed. Also, because air is a bit denser in the winter than in the summer, you can expect greater output from a wind turbine in the winter — about 13 percent more. This is especially important to those who are installing hybrid wind/PV systems.

Although you can't control air density, you *can* control swept area and wind velocity. Swept area, as explained earlier, is determined by rotor size. The bigger the rotor, the greater the swept area. Select a wind turbine with the largest possible swept area.

You can also affect wind speed by choosing the best site and installing your wind turbine on the tallest tower possible. This allows you to raise the wind turbine into the fastest winds and will have an enormous effect on electrical production. For example, by doubling the wind speed from 8 miles per hour to 16 miles per hour, power production increases by 800 percent!

Guyed Towers

Guyed towers typically consist of steel pipe or lattice uprights supported by cables called *guy wires*, a.k.a. *guy cables*, that run from the tower to anchors in the ground (Figure 9-10). (Please note that it is guy wires or cables, not *guide* wires.) Guyed towers are the cheapest towers and are therefore very widely used in the small wind industry. Steel pipe can be used for the masts of all household-sized wind turbines. The steel pipes are assembled section by section, secured by bolts or slipped together, then erected using a crane.

For larger small wind turbines, the lattice tower is often the tower of choice. They're strong, mass-produced for the telecommunications industry and therefore widely available, and they are cheaper than other options. They're also available in different strengths. Like free-standing lattice towers, they can be assembled in the horizontal position on the ground then

BERGEY WINDPOWER

Fig. 9-10:
*Guy wires help
hold many
towers in
place. Wires
are installed in
groups of three
to four. Typically,
each segment
of the tower is
guyed.*

radius — that is, how far out the anchors for the guy wires need to be for optimal strength.

Tower Kits

If all of this seems a bit daunting, don't be dismayed. A professional installer will design a system for you, order the components, assemble the tower, mount your wind turbine, and raise the tower ... for a price, of course. If you'd like to try your hand at this, you'll be happy to know that several manufacturers sell tower kits designed and engineered specifically for their wind turbines. Southwest Windpower, Lake Michigan Wind and Sun, Xzeres (formerly Abundant Renewable Energy), and Bergey, for instance, all sell tower kits. Because it doesn't make sense to ship fairly inexpensive, but very heavy and bulky steel tubing long distances, manufacturers typically provide all of the materials you need *except* the pipe. These kits may not include the anchors because the type of anchor varies from one location to another, depending on the soil type. You can purchase steel pipe locally at a steel and pipe supplier. You can purchase the anchors from the tower kit manufacturer, although most anchors are made of steel-reinforced concrete poured on site. To learn more about towers, you can read my book, *Power from the Wind,* or attend a workshop on small wind energy systems such

tilted into place, or they can be assembled upright, one section at a time or in their entirety, using a crane.

In all guyed towers, strong steel cable is attached to the tower and anchored in the ground using an assortment of anchoring mechanisms, depending on the soil type. Stranded wire cable ranging from one quarter to one half inch is typically used.

Installers use three guy wires for each section of the tower, located 120 degrees apart. Manufacturers typically specify guy as those I offer at The Evergreen Institute in Missouri.

Tilt-Up Towers

The third type of tower is the guyed tilt-up tower. Unlike free-standing and fixed guyed towers, guyed tilt-up towers can be raised and lowered to perform inspection, maintenance, and repair.

Tilt-up, guyed towers come in many heights — up to around 130 feet (40 meters). They are most often made from steel pipe, although guyed tilt-up lattice towers are also available. Guyed tilt-up towers require four sets of guy cables every twenty to forty feet, rather than three in fixed guyed towers. In addition, the guy cables are located 90 degrees apart. As in the fixed, guyed tower, guy cables of guyed tilt-up towers provide support. The fourth set of cables, however, enables workers to safely raise and lower the tower. Without the fourth set of cables, the tower would topple as it was tilted into place or lowered to the ground for servicing.

As illustrated in Figure 9-11, a tilt-up tower is raised and lowered with the aid of a gin pole. This steel pipe (or a section of lattice tower) is usually permanently attached to the base of the tower at a right angle (a 90° angle) to the mast. It serves as a lever arm that allows the tower to be tilted up and down for inspection, maintenance, and repair.

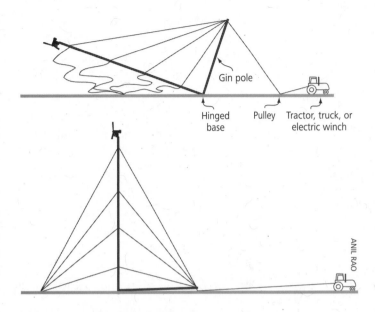

Fig. 9-11: *The gin pole shown here allows an installer to tip a large tower into place without an expensive crane.*

Fig. 9-12: *The gin pole and mast of the tower are connected to a hinge, which permits the tower to tilt up and down.*

As shown in Figure 9-12, the gin pole is attached to the base of the tower which is attached to the hinged joint. When the tower is down — that is, lying on the ground and ready to be raised — the gin pole sticks straight up. When the tower is vertical, the gin pole lies near and parallel to the ground. As illustrated in Figure 9-11, a steel cable connects the free end of the gin pole to a lifting device such as a tractor or better yet, an electric winch, which is used to raise and lower the tower (discussed shortly).

Tips on Finding a Supplier or Installer

If wind is a viable option for you — that is, you have enough wind in your area and can afford to buy and install a turbine — you have two obvious options for proceeding. You can purchase and install the equipment yourself, or you can hire a professional. If you select the first route, you'll need to do a lot more in-depth research. You'll also need to read up on installation. You'd also be wise to take a course on wind energy that includes hands-on installation work. A couple of installations would be even better. Be careful. Installation can be dangerous to people and equipment. You don't want a $5,000 to $25,000 wind turbine to come crashing down because you made a mistake during installation.

When purchasing a wind turbine, be sure to check out warranties and read the fine print. Most warranties are for five years, although a few manufacturers recently boosted theirs to 10 years. Also check out how long the companies you are considering have been in business.

The second option is to go with a local supplier/installer. Their expertise can be highly valuable. A local expert can also help you troubleshoot problems once the system is up and running and can repair and maintain your system, if need be.

Yes, you read that right.

Wind energy systems need periodic maintenance — like any mechanical and electrical system. Wind turbines and towers, for example, must be inspected periodically — usually twice a year. If problems are encountered, they must be corrected. Some larger wind turbines may even need grease or oil changes. Battery-based systems require even more maintenance, a topic briefly described in Chapter 8 and more thoroughly covered in my book, *Power from the Wind*. If you are not into maintenance, don't buy a wind turbine.

Manufacturers have made tremendous strides in improving the quality of their wind generators since the early 1980s, mostly by simplifying the turbines and using stronger, more durable materials, such as carbon-reinforced blades and new metal alloys. However, don't forget: wind turbines are exposed to extreme weather and amazing stresses (heat and cold, snow, ice, and fierce wind). Fortunately, many problems that arise are simple ones that can be fixed in seconds, like loose bolts. If left unattended, though, small problems can lead to catastrophic failures.

Even blades require maintenance. Some need to be repainted from time to time. Protective tape on the leading edges of blades protects them against grit and insects in the air. The tape will need to be replaced every so often. And, of course, bearings will eventually wear out and need replacement.

To maintain a wind generator, you'll either need to periodically ascend the tower using a climbing harness — very carefully, so as not to fall — or you'll need to lower the turbine to ground level (the main reason for installing a tilt-up tower). After 20 years of successful operation, the guts of the wind turbine may even have to be ripped out and replaced. As Mick Sagrillo points out, "The life of a wind generator is directly related to the owner's involvement with the system and its maintenance." If you expect a wind turbine to work doggedly in its harsh environment without maintenance, wind power is not for you. That's a pretty unrealistic expectation.

Some Tips on Installing a Wind Turbine

To be effective, wind turbines need to be installed in a good, windy site, usually upwind of buildings and other obstructions. If installed downwind, turbines need to be significantly past the wind shadow — the turbulent eddy created by obstructions. The best wind, as already noted, is the smoothest and strongest wind, which is 80 feet or more above the ground. The higher, the better. To be effective, a wind turbine typically needs to be erected on a tower at least 80 to 120 feet above ground level to access the best wind and, as already noted, the entire rotor must be at least 30 feet above the closest obstruction or treeline within 500 feet. Whatever you do, do not mount a wind turbine on a roof or against a building — even if a supplier provides special mounts for such applications. The vibrations will be conducted into the building and are very annoying. Sadly, I know this to be true because I tried it … just to test it, of course. (Yeah, right!) Moreover, if the turbine is on or against a building, the building itself will create air turbulence that will lower performance and increase wear and tear on the turbine.

While we're on this topic, be sure to mount your wind generator as close to the inverter as possible. This is important for two reasons: to reduce line loss and to reduce the need for larger (and more costly) electrical wire. Installation manuals will help you determine the gauge wire you will need.

Before you decide to install your own wind generator, be sure to read more on the subject.

FINANCIAL MATTERS

Like a good salesperson, I've given you the scoop on wind power before I drop the cost bomb. How much does it cost to install a wind turbine?

As noted earlier, wind generators for household use vary from the smallest 1,000-watt turbines to the largest 20,000-watt unit. If you simply want to supplement your electrical energy from the grid, a wind turbine of 1,000- to 3,000-watt rated output might be ideal. When buying, look for models in this range, then compare them using swept area, weight, AEO, and costs. A 1,000- to 3,000-watt wind generator might also work if you live in a tiny cabin, or you are extremely efficient. And it might suffice if you are installing a hybrid system — a combination of wind and PVs — or some other renewable resource such as microhydro (see Chapter 10).

If you want to go off-grid completely and are hoping to supply electricity solely from a wind turbine, you'll very likely need a larger wind generator — a turbine with a rated power output of more than 3,000 watts — possibly 6,000 to 10,000 watts. It all depends on how much energy you need. Again, whatever you do, be sure to make your home as efficient as possible in its use

of electricity first. Small investments in electrical efficiency pay huge dividends by reducing system size and costs.

Also, remember that you get what you pay for. Heavy-duty wind turbines generally cost more but are more durable and will very likely outlast their cheaper, lighter, high-speed cousins. A good heavy-weight wind turbine will very likely last 20 years — or more. Its lightweight cousin will last half that time, maybe even only one fourth as long, according to Sagrillo. So, savings on a less expensive model are eaten up by replacement costs.

Another important consideration when designing a wind system is that tower height makes a huge difference in power output. Because power increases dramatically with wind speed and because wind speed increases with height, raising the tower a bit can greatly boost a system's output. So rather than paying twice as much to purchase a more powerful wind turbine, consider a smaller investment in tower height to increase energy production.

Wind turbine costs vary tremendously. Wind turbines around 1,000 watts rated output cost about $2,500 with a charge controller, although there are some models that cost more than $5,000. Wind turbines ranging from 2,500 to 6,500 watts with charge controllers will run about $9,000 to $20,000. Costs for the largest residential wind turbines — 8,500 to 20,000 watts rated output — are $20,000 to $40,000. But don't forget the tower, another big expense. Tower prices vary considerably, depending on the height of the tower, type of tower (monopole vs. guyed tilt-up, for instance), and type of material. The most expensive is the free-standing monopole, followed by the free-standing lattice, followed by the tilt-up. The least expensive is the guyed lattice tower. You can easily pay $12,000 to $25,000 for a tower and installation, sometimes more, for a 120-foot tower.

When calculating costs, remember that in the United States the federal government provides a 30 percent federal tax credit for wind systems. The USDA offers a 25 percent grant for wind systems for rural businesses. Very few states and utilities, however, offer financial incentives. Two exceptions are New York State and Wisconsin.

WIND POWER WITHOUT INSTALLING A WIND GENERATOR

If a wind generator is not right for you, don't fret. Many utilities offer green power to their customers. Green power is electricity generated from renewable sources, mostly gigantic wind turbines on commercial wind farms, like those shown in Figure 9-13. Green power may be produced by other renewable sources as well, but wind is the dominant source. It's the cheapest and most widely developed commercial renewable energy source in the world, other than hydropower from hydroelectric plants on rivers. In Colorado, the local utility offers its customers an option to purchase blocks of wind energy from ever-expanding wind farms. Customers can opt to buy as many blocks as they want, provided the company has enough to sell.

Even if the utility company in your area does not have its own wind farm, it may currently be purchasing power from renewable producers and can offer you green energy just the same. At this writing, utilities in at least 36 states in the United States offer some form of green power. What that means is that many US households could choose some type of green power directly from their local utility or through the competitive marketplace. California and Pennsylvania have been the most active markets for green power. (See Blair Swezey and Lori Bird's article, "Businesses Lead the Green Power Charge" in *Solar Today*.)

Chances are you've heard about green power and already know whether it is available. If not, ask your local utility. As the markets for electricity deregulate, more and more companies will be able to provide power on electrical grids owned by others; you may be able to purchase green power from one of these competitors. Even if it costs a few bucks more each month — and it usually will — it's well worth it.

If there are no utilities in your area that sell green power, you can still play an active role in this growing industry by purchasing a green tag. A green tag is a small

DAN CHIRAS

Fig. 9-13: *This large wind turbine in northeastern Colorado is used to create green power for a local municipality.*

subsidy to power companies producing green power. It supports their green power programs. Bear in mind, however, that you won't actually receive the electricity. Someone else will, but you will help make it happen. Green tags are sold by a number of companies and go by different names. The Los Angeles Department of Water and Power, for example, calls their green tags "Green Power Certificates." Pacific Gas and Electric sells "Pure Wind Certificates." Waverly Light and Power in Iowa sells "Iowa Energy Tags." Big city governments, such as those of Chicago and San Francisco and key federal agencies, such as the EPA, the DOE, and the US Postal Service, are currently making substantial commitments to buy green power. With these advances, it seems very likely that homeowners are going to have ever greater opportunities to purchase green power for themselves.

At the present time, electricity from clean, reliable sources costs a little more; on the other hand, some service charges may be deducted from your bill, reducing the costs. In addition, the cost of wind power is likely to fall in the very near future as more wind turbines go on line and more improvements are made in wind generator technology. Furthermore, we've recently seen that while prices for electricity from coal have increased, wind-produced electricity has remained the same.

Remember: good planets are mighty hard to come by. A small investment in environmentally friendly electricity is a small price to pay to create a sustainable way of life and a better future. Investments in renewable energy will also help build stronger economies that resist the potential turmoil caused by declining supplies of conventional fuels.

The Cost of Green Power

Over time, as wind-generated energy becomes more widespread, the price is bound to plummet. Newer turbines are already producing electricity more cheaply than natural gas — at a price only slightly higher than coal.

CHAPTER 10

MICROHYDRO

GENERATING ELECTRICITY FROM RUNNING WATER

Humans have been tapping the power of flowing water for centuries. Early New Englanders, for instance, tapped the power of water flowing in the streams that ran through their towns by installing small dams and water wheels. The water wheels powered the machinery of textile factories and grain mills where wheat was ground into flour used to make bread, a staple of early American life. Hydropower continues to play a pivotal role throughout the world today. Its primary value, however, is as a source of electricity. Today, hydroelectricity constitutes 7 percent of the total electrical generation in the United States, and 21 percent in Canada.

But that's not hydropower's only claim to fame.

Hydropower is not only a significant source of energy, it also has the distinction of ranking number one in annual energy production among *all renewable energy resources.* That is, of all renewable energy sources in the world, it's the "top dog." Large-scale hydropower projects, however, are not the main topic of this chapter. Rather, we'll focus our attention on microhydro — small-scale systems that provide electricity to homes, small farms, ranches, and small businesses.

Yes, even small businesses.

One of the best examples of a business application is the shop run by Don Harris in Santa Cruz, California. Where this pioneer in renewable energy manufactures turbines for microhydro systems. According to Scott Davis, author of *Microhydro: Clean*

Fig. 9-1:
Don Harris,
a pioneer in
microhydro
power in North
America,
manufactures
microhydro
turbines/
generators in
his shop in the
hills outside of
Santa Cruz, quite
fittingly powered
by hydropower.

Power from Water, "Even at the smallest of scales, water power continues to be a reliable and cost-effective way to generate electrical power with renewable technology." Although microhydro is a reliable and economical source of electricity for those wishing to go part way or entirely off-grid, it does have its limitations. First and foremost, you need a stream or river nearby — one that has a consistent and sufficient flow and offers sufficient water pressure (more on this shortly) to produce enough electricity to make the investment worthwhile.

Unfortunately, there aren't that many microhydro sites available worldwide into which homeowners can tap. For this reason, microhydro is of limited use in most countries. However, for those who are lucky enough to live near a suitable stream — usually those living in mountainous or hilly terrain — microhydro offers great promise. Even people in the flatlands can generate electricity from water if there's a large enough stream nearby that flows at a sufficient rate of speed. It is for these lucky individuals — and for folks who just want to learn more about this fascinating energy source — that this chapter was written.

AN INTRODUCTION TO HYDROELECTRIC SYSTEMS

Hydroelectricity is based on some pretty simple concepts. If you have read the wind power chapter in this book or have studied conventional electrical production by coal, nuclear, or geothermal energy, you'll see the similarities instantly. In a microhydro system, moving water turns a turbine. The turbine spins a generator. The generator (or alternator) produces electricity. These components are common to all of the electrical generating equipment discussed in this book, except PV modules.

"Many other components may be in a system, but it all begins with the energy ... within the moving water," says Dan New, author of "Intro to Hydropower" published in *Home Power* magazine, Issues 103–105. (This information can also be found on

DON HARRIS

Dan New's website canyonhydro.com under "Residential Systems." Click on the "Guide to Hydropower.") The energy of moving water is produced by gravity, that magical force that propels water down the slightest of gradients. It is important, however, to point out that water reached the heights from which it flows downward thanks to two other natural forces — evaporation and precipitation.

Water evaporation is triggered by solar energy. Sunlight striking the Earth and various water bodies causes moisture to evaporate, and that is where the flow of water through the hydrological cycle begins. You might say that it is all downhill from that point on. Because of this, hydropower is just another form of solar energy. That mysterious force, gravity, plays its part, too.

As in other renewable energy systems, microhydro systems fall into three broad categories:

(1) off-grid, (2) grid-connected, and (3) grid-connected with battery backup. (These distinctions are outlined in Chapter 8 for those who aren't familiar with the terms.)

THE ANATOMY OF A MICROHYDRO SYSTEM

Microhydro systems are electrical generating systems for use on a small scale, usually for residential power in remote mountainous terrain. Microhydro is usually "installed" along small streams or rivers close to the buildings where it will be used — the closer the better! To protect the stream and those creatures that depend on it, microhydro systems generally divert only a small portion of the current from the waterway. This water temporarily borrowed from the stream is diverted into a pipe or specially built channel or canal (that typically runs alongside the waterway) to a turbine some distance below the water intake. The turbine is connected to an electrical generator, described shortly.

Microhydro systems exist in two basic configurations: low-head and high-head. (Head refers to water pressure created by the vertical distance water flows, as you shall soon see.) Most homeowners install high-head systems.

The Renewable Energy of Choice

For those who are fortunate enough to have a good site, hydro is really the renewable energy of choice. System component costs are much lower and watts per dollar return is much greater for hydro than for any other renewable source. Microhydro, given the right site, can cost as little as a tenth of a PV system of comparable output. Moreover, hydropower users often are able to run energy-consumptive appliances that would bankrupt a PV-system owner, like large side-by-side refrigerators and electric space heaters.

— John Schaeffer, *Solar Living Source Book*

Fig. 10-2:
High-head
microhydro
systems consist
of a water
intake structure,
a pipeline or
penstock, a
turbine/generator,
and an outlet.

High-Head Microhydro

Figure 10-2 shows the basic components of the most widely used microhydro system, the high-head system. As illustrated, this system consists of a specially built water intake structure. It is constructed in a stream or river, usually along its banks, so as to minimize disturbance to the aquatic environment. One of the key components of the water intake structure is a small settling basin that allows grit to precipitate out. (Grit and silt in water can wear out the moving parts of the microhydro turbine.) Another key component of the water intake structure is a screen that removes debris such as leaves and branches that can clog the pipes and the spray nozzles on the turbine. They can also damage the turbine runner (the wheel inside the turbine that spins when struck by water). Water typically flows from the intake structure into a pipeline (called a "penstock" by professionals). The pipe typically runs alongside the river or stream — not in the watercourse itself — to protect the pipe from raging flood waters that can rip it apart. The pipe carries the water downhill to the turbine, usually located in a simple shelter — either a small, sturdy, waterproof vault or a small shed. Water flowing through the turbine causes a wheel (the "runner") inside the device to turn (Figure 10-3). A steel shaft connects the runner to the generator. When the runner and shaft spin, they cause the inner workings of the generator to spin, producing electricity. (Because the generator in a microhydro system operates like the generators in wind turbines, you might want to check that section out now. See page 243)

Water flowing through the turbine housing flows back into the stream. This return flow is referred to as the *tailwater*. As noted earlier, microhydro systems

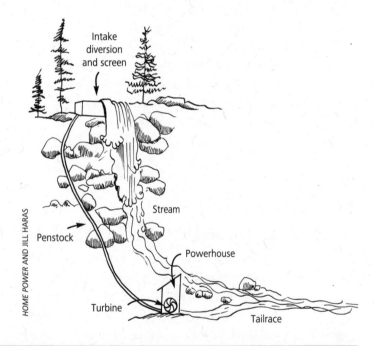

HOME POWER AND JILL HARAS

Intake diversion and screen

Stream

Penstock

Powerhouse

Turbine

Tailrace

AC and DC Microhydro

Microhydro systems produce both DC and AC power, depending on the design. However, DC systems are by far the most commonly used in small-scale applications.

produce DC and AC electricity depending on the design of the generator. As you may recall from reading previous chapters on solar electric and wind systems, AC electricity powers our homes and offices. As a result, DC electricity produced by the generator in a microhydro system must first be sent to an inverter. It converts the DC electricity produced by the generator to AC power.

The inverter also boosts the voltage of the current from the generator, typically from 12, 24, or 48 volts, depending on the model, to 120/240 volts, which is standard household current.

Low-Head Microhydro

Low-head microhydro systems are a bit simpler than the high-head systems just described. As illustrated in Figure 10-4, they require a screened intake or a small dam across part of a stream or river with a settling basin to allow grit and silt to precipitate out. The intake structure empties into a fairly short diversion canal that delivers water to a vertical draft tube often only ten feet long. A draft tube sends diverted water back into the stream and is where the turbine is located. The water flowing through the system turns a propeller-like turbine. It is connected to a generator that produces electricity. Water is returned directly to the stream. Such systems are often built near small water falls.

AC Microhydro Systems

AC microhydro systems produce a form of electricity that can be used directly without conversion. In off-grid systems, the AC electricity will, however, need to be converted back to DC to be stored in a battery bank.

HARRIS HYDROELECTRIC

SCOTT DAVIS

Fig. 10-3: *(a) Pelton and (b) Turgo runners.*

ASSESSING THE FEASIBILITY OF YOUR SITE

Before you decide to purchase a microhydro system, you — or better yet, a professional microhydro installer — will

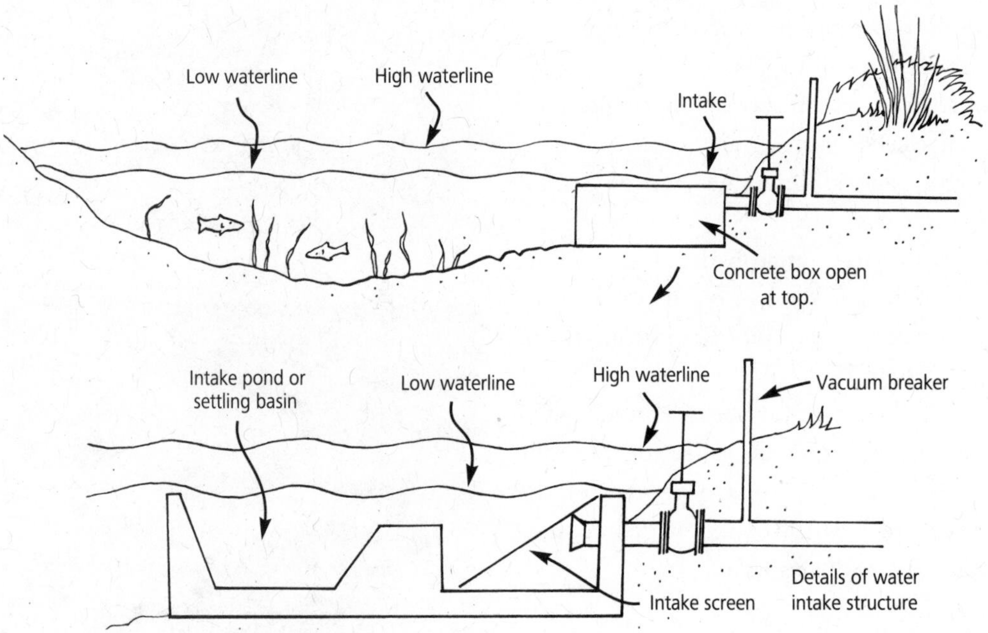

Fig. 10-4:
A typical water intake structure requires a stilling basin to allow sediment to precipitate out. It also requires a deep pool so that the pipe remains free from ice.

need to study your site very carefully to determine whether a system could work and how much electricity it could generate.

Analyzing Your Electrical Energy Needs

Before you assess the electrical generating capacity of a stream, however, it is important, indeed essential, to analyze your electrical energy needs. This process is described in Chapter 8, so I won't repeat it here, other than to offer a few reminders. First and foremost, when determining your electrical demand, it is important to be accurate — and generous — in your assessment. Accuracy is vital to your success, but it is often difficult to achieve.

Generous estimates are important because most people use a lot more electricity than they think they do.

Why?

When estimating electrical consumption for a new home, people often begin by listing all of the components of their load, that is, their household electrical demand. This includes appliances, lights, fans, pumps, coffee makers, aquariums, and the assortment of other electronic devices they have. They then list how much power each electronic device consumes, and how long the device is used each day.

The problem with this approach is that very few people really know how long they run the TV or the toaster or the microwave each day. So estimates can be pretty inaccurate. If you are installing a system for an existing home or work space, your task is considerably easier: you simply consult your energy bills for the past two or three years to calculate monthly electrical consumption. If you haven't saved your bills, you can contact your local utility company. They will gladly provide records. Armed with this data, you can arrive at a pretty accurate estimate of the electrical energy demands that your system will need to meet. But what do you do if you are buying a house from someone else? Get on the phone and call the utility and ask for previous utility bills. Bear in mind, however, that previous utility bills, while accurate, represent use by a different household, and may differ dramatically from your use of energy. The previous occupants may have been more miserly or much more extravagant than you — or somewhere in between. Their energy consumption, in short, could be quite different from yours. Before you arrive at a final estimate of electrical consumption, however, remember to trim your own electrical consumption. In so doing, you can dramatically reduce the cost of a renewable energy system. Changes in habits (like shutting off lights when you leave a room) and energy efficiency measures (compact fluorescent lightbulbs and LED lightbulbs, for instance) can dramatically reduce your monthly electrical consumption. Remember, as I've said many times before, efficiency is *the* most cost-effective means of supplying power to a home — or any building, for that matter.

And, lest you forget, the more efficient you are, the smaller your system requirements. The smaller your system requirements, the less money you'll have to spend up front.

ASSESSING THE POTENTIAL OF A MICROHYDRO SITE

Once you have determined your family's electrical energy demands, it is time to assess the potential of your site. You'll need to determine how much electricity a microhydro system could produce and the percentage of your needs that it could meet. Before you do this, you should be aware of the difference between continuous power and daily power production and use.

Unlike other renewable energy systems, microhydro systems can produce electricity 24 hours a day, 7 days a week, 365 days a year. A system, for instance, might produce 200 watts of continuous power day in and day out. So, over a 24-hour period, this system would produce 4,800 watts-hours, or 4.8 kWh. To determine monthly output, multiply the daily output by 30. That would give you 144,000 watt-hours per month (144 kilowatt-hours). If the monthly electrical energy production of the system matches your household need, you'll be in fine shape. However, most conventional homes — homes that don't practice energy conservation — consume 25 kWh per day, or a little over 750 kWh per month. So, the system we are looking at would fall short of most families' needs. If you are powering a home that is connected to the grid, use the annual electrical energy consumption to size your system. If you are powering an off-grid home, you'll need to determine the month with the highest electrical demand, and size the system so that it meets that demand — or find some other way to supplement production, such as PVs or a small wind-electric system.

Although a microhydro system produces a fair amount of energy over a 24-hour period, what happens if your demand exceeds its output at any one time? For example, suppose your system produces 200 watts of continuous power. Although that may be enough in a given day to meet all of your household needs, what happens if you need 1,000 watts of power to run a microwave, 100 watts of power to run some lights and your computer, and 50 watts to run the television — all at the same time? And what happens when your washing machine or a power tool is turned on? As described in Chapter 8, power tools and household appliances such as dishwashers and washing machines require more power to start than to run — quite a lot more. This is known as *surge power*.

As Scott Davis points out in *Microhydro: Clean Power from Water*, "It is entirely possible to have a system that has plenty of kilowatt-hours available, but not enough capacity to start critical loads" or to meet demands that exceed continuous power output.

Demand in excess of continuous power output can be met by grid power or battery backup. In a grid-connected system, for instance, additional electrical power required to satisfy excess loads, including surge loads, is derived from the utility line connected to your home. In homes equipped with battery backup, electricity required to meet electrical demands that exceed the continuous output of the microhydro system comes from the battery bank. It is stored there during periods of excess.

Peak power demand can also be "managed away." That is, homeowners can avoid power consumption in excess of the continuous output of a system by managing their load better — spreading out use so as not to exceed a system's continuous power output. Peak loads can also be satisfied by using a small gas, diesel or propane electrical generator. Assessments of the site's electrical energy production compared with household energy consumption help homeowners determine whether a site will produce enough electricity on a daily basis to meet their household needs. Understanding demand patterns, however, are also necessary. This will help homeowners decide what type of system should be installed and whether battery backup or a grid connection is needed. I will cover this in more detail shortly, but first let's take a look at ways to assess the potential of a site.

Assessing Head and Flow

Rather than go into minute detail on the ways you determine the potential of a microhydro site, I will give you a general overview of the process. If you want to install a system, you will need more information than I could possibly supply here. You should also consult one or all of the following resources: (1) Dan New's article in *Home Power* magazine, "Intro to Hydropower. Part 2: Measuring Head and Flow;" (2) Scott Davis' book, *Microhydro: Clean Power from Water*; and (3) *Residential Microhydro Power with Don Harris*, a video produced by Scott Andrews (you can purchase a copy through Gaiam Real Goods).

When assessing the potential of a site, you need to determine two factors: head and flow. In the words of Dan New, "You simply cannot move forward without these measurements." Your site's head and flow will determine everything about your hydro system — pipeline size, turbine type, rotational speed, and generator size. Even rough cost estimates will be impossible until you've measured head and flow. New advises, "When measuring head and flow, keep in mind that accuracy is important. Inaccurate measurements can result in a hydro system designed to the wrong specs, and one that produces less electricity at a greater expense." You may want to consult a professional with lots of experience in microhydro to help out or to make accurate measurements. If you're lucky enough to have a local installer, it's not a bad idea to bring him or her in on the project. It may cost you more, but you could end up with a much better system — and more energy — than if you did all of this yourself.

Measuring Head

Head is an engineering term for water pressure created; in the case of microhydro systems, it is created by the difference in elevation between the water intake and the turbine. Head is usually measured in feet or meters when assessing site potential. The difference in elevation between the intake and turbine can be measured by a variety of methods. For example, you can use an altimeter on a wrist watch (although cheaper ones can be off by 150 feet or more). You can also use a topographical map of your site to determine the differences in altitude between the intake and turbine.

While both of these methods work, they're often subject to error. As Scott Davis points out, "The accuracy of the power estimate and jet sizing ... requires measurement of head within 5 percent or so." As a result, experts typically recommend direct measurement. The most accurate ways to measure vertical distances rely on some simple tools — a transit or a level (either a carpenter's level or a hand level) and a calibrated pole, for example, an eight-foot length of pole, marked off in feet and inches. You can even get by with a level and two people. One method is shown in Figure 10-5.

Rather than try to explain each of the methods commonly used to measure head, I recommend that you read Dan New's piece in *Home Power* magazine or check out Scott Davis' book, mentioned earlier. They'll walk you through the techniques.

(By the way, New's method for measuring uphill is one of the simplest you will find anywhere.) Don Harris shows the technique in his video, too. The various techniques for measuring head allow you and an assistant to work your way up or down a hill, measuring the drop (head) from one spot to another. When done, you simply add together all of the measurements for head from the water intake to the turbine to determine the total head.

Head can also be measured using garden hose or flexible tubing and a pressure meter. Dan New explains this technique, but points out that it works best for short distances — those in which a hose can be run from the proposed intake to the proposed turbine location. If the distance is greater than the hose length, you can link several hoses together, as long as the connections are water tight. Or you can move the hose from one spot to another, taking multiple measurements. The individual pressure settings are then added together

Fig. 10-5: Head is the vertical drop between the intake and turbine. It is measured in a number of ways. This drawing shows one convenient method for measuring head.

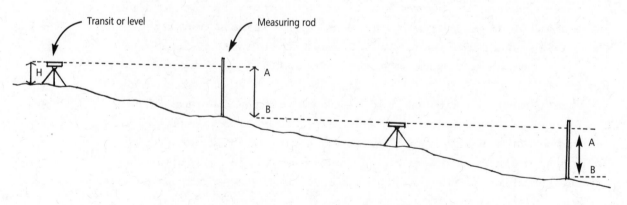

to get the total head. In this case, however, head will be measured in water pressure, pounds per square inch or Newtons per square meter (with metric system meters). Read Dan New's piece before you tackle this project, as there are ways that you can get false readings.

Adding all of the measurements you've carefully taken yields the true vertical distance from inlet to turbine, something that the microhydro folks call the "gross head" for a proposed site. But that's not sufficient for calculations. You need to convert gross head to net head. Gross head is the pressure (measured in feet or meters) in the system when the water is not flowing. It is static head. It is the pressure reading you'd get if you ran a pipe from inlet to turbine, filled the pipe with water, but capped the end of the pipe so water could not flow out. If you opened the pipe and let water flow through it, you'd see a drop in pressure. That's the net head.

What accounts for this difference?

Net head, the measurement you get when water flows through the pipe, is less than gross head because of friction. Net head is less than gross head because friction slows the flow of water through the pipe. That is to say, water flowing through the pipe loses some energy due to friction losses — water molecules interacting with the interior surface of the pipe. As a result, net head is always less than gross head.

Before you can determine the friction loss, you will need to know the pipe length and the flow rate — how much water will flow through the pipe. I explain below how flow is measured. For now, it is important to note that the larger the pipe, the less friction loss. Conversely, the smaller the pipe, the more friction loss. Net head declines substantially in smaller diameter pipe.

But that's not all.

Friction losses also increase at higher flow rates. The faster the water flows through the pipe, the more friction. Moreover, friction losses increase in longer pipelines. Bends in a pipe also increase friction losses that decrease head.

These four factors — diameter of the pipe, flow rate, length of the pipe, and bends in the pipe — need to be considered very carefully when designing a system.

As a general rule, "a properly designed pipeline will yield a net head of 85 to 90 percent of the gross head you measured," says New. Greater losses will result in unacceptable reductions in electrical production. But like anything, there are exceptions and nuances to be considered. As New points out, "higher losses may be acceptable for high-head sites (100 feet plus) but pipeline friction losses should be minimized for low-head sites." While larger pipes reduce head loss resulting from friction, they cost more; so there's often a tradeoff between cost and performance. As a general rule, you should design your system in such a way that head loss is no more than 10 to 15 percent. Before you can calculate head loss, you need to determine flow rate in your stream or river.

Measuring Flow Year Round

Stream flow can be measured in one of several ways. For best results, stream flow should be measured during different times of the year because fluctuations are common in perennial streams (those that flow year round). Even in temperate rain forests like those of the Pacific Northwest, there's a dry season during which stream flows are diminished. In mountainous areas, like the Sierras and the Rocky Mountains, perennial streams flow year round, but stream flows peak during the spring months as a result of spring snowmelts. Flows decline dramatically in the summer, fall, and winter. So, to determine how much power a system can generate, you need to know a stream's high-, medium-, and low-flow rates.

Flows in perennial streams can be reduced to a mere trickle, forcing homeowners to turn to other sources of electrical energy. Take the case of John Schaeffer, who founded Real Goods and the Solar Living Institute. John and his family live off-grid in northern California where they tap into the power of flowing water in a stream on their beautiful property. In this region of the country, rain is plentiful in the winter, but declines dramatically once June rolls around. As a result, the stream on the Schaeffer's property pretty much dries up in the summer and stays dry until late fall. (I know, I've seen it!) As a result, he and his family get much of their electricity from their microhydro power system during the late fall and winter and early spring when it is quite rainy. During the summer and early fall, their PV system makes up the difference, supplying virtually all of the electricity they need. So, remember to measure flow rates, if you can, during the spring, summer, winter, and fall for the most accurate assessment of a site. Remember, too, that in climates where freezing occurs, water continues to flow in streams under the protection of a layer of ice and snow. You can tap into this energy source even though the stream appears to be frozen over for the winter. It's wise, however — and environmentally responsible — not to take all of the water from the stream then or any time of the year!

Measuring Flow Rates

Stream flow can be measured by one of three methods. The simplest is the container fill method (Figure 10-6). This is particularly useful in really narrow streams. It requires a large bucket and a stop watch. To make a measurement, you simply find a narrow spot in the waterway where you can fill a single bucket. A bucket is then thrust into the spot and filled by one person. Someone else uses a stopwatch to time how long it takes the bucket to fill. If the five-gallon bucket fills in 15 seconds, you would multiply five gallons by four to determine the amount of flow in a minute. In this instance, the flow rate would be 20 gallons per minute.

The second method is the float method, well described in Dan New's piece in *Home Power*. This technique is useful for larger streams but in order to do use it, you will need to locate a fairly straight section of stream — one that's about ten feet long, as shown in Figure 10-7. In the float method, you first measure how long it takes for a floating object to travel ten feet. Let's say the object — a tennis ball, for instance — covers the ten feet in five seconds. This figure will then be used to calculate flow. But before you can do so, you need to know the average depth across a section of the stream, as shown in Figure 10-7. Depth measurements are taken every foot across a section of the river. The measurements are all added together, then divided by the number of measurements, yielding average depth. This figure is then multiplied by the width of the stream. If the stream is 6 feet wide and the average depth measurement is 1.5 feet, the cross-sectional area is 9 square feet. If the floating object takes 5 seconds to travel 10 feet, the stream is flowing at 2

Fig. 10-6: *The container fill method is one way to determine flow rates in small streams.*

SCOTT DAVIS

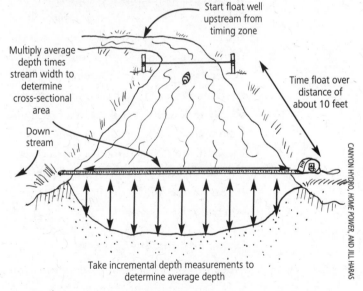

Start float well upstream from timing zone

Multiply average depth times stream width to determine cross-sectional area

Down-stream

Time float over distance of about 10 feet

Take incremental depth measurements to determine average depth

CANYON HYDRO, HOME POWER, AND JILL HARAS

Fig. 10-7: *The float method, described in the text, is another way of determining flow rates of streams and rivers.*

feet per second or 120 feet per minute (2 feet per second x 60 seconds per minute = 120 feet per minute). If the cross-sectional area is 9 square feet, you simply multiply 120 feet per minute by 9 square feet and the result is 1,080 cubic feet per minute. That's the rate of flow. (This needs to be adjusted to take into account frictional losses along the bottom of the stream.) Multiplying 1,080 cubic feet per minute by 0.83 yields the actual flow.

The next method, known as the weir method, is even more involved, but it's the most accurate of the techniques. This method requires the construction of a temporary dam across the river. Because this technique is more complicated, and involves reference to a weir table, I'll let those interested in learning more about the method read Dan New's account in *Home Power*, Issue 104. Basically, what you need to know is that to calculate the flow, you simply measure water level in a rectangular opening in the weir. You then refer to the weir table that provides data on flow in cubic feet per minute. You multiply this figure by the width of the opening, and you're done.

Protecting the Environment

Once you have determined flow rate, you need to determine how much water can be removed from the stream without disrupting the aquatic ecosystem. Obviously, the less water that's removed, the less impact you'll have. To make an assessment will require more training than most readers possess. I recommend that you consult an aquatic biologist for a site assessment and recommendations. Your state division of wildlife may be able to help you out. Local permitting agencies may also have guidelines or may supply personnel. A professional stream biologist may be able to give you the best information about the potential impact of damming up part of the river and/or removing water from the stream. Renewable energy is clean and great for the environment, but you don't want to trash your stream to produce clean power. So be careful, and show respect for the natural world that gives so much to us.

What's the Next Step?

After determining head and water flow, you need to measure the length of pipeline. You can measure the length by pacing off the distance from the inlet to the turbine; however, it is best to measure it directly, using a 100-foot tape measure. You'll also need to measure the distance from the turbine to your house and the battery bank. As in other renewable energy systems that produce DC power, it is best to keep the batteries as close as possible to the source of electricity (in this case, the turbine). Low-voltage DC electricity doesn't travel well, and longer distances require larger diameter wire that is much more expensive than standard 12-gauge electrical wire. This is one reason why it is a good idea to install an AC generator in a microhydro system. The higher voltage AC travels long distances with low losses. It is also wise to keep the batteries warm year round. Lead acid batteries, the type most commonly used in these and other renewable energy systems, perform optimally at temperatures of around 70°F (21°C). (For more on batteries, see Chapter 8.)

With this information, you can begin to design a system. You'll need to pick out a turbine that suits your purpose, and then determine pipe size. You can make these determinations yourself, or you can contact a local supplier. National suppliers of microhydro equipment, such as Gaiam Real Goods, can also lend a hand. Their staff can assist you in sizing and designing a system. They'll provide a worksheet that asks for data that is used to generate a design for your system via a computer-sizing program. Their computer program allows them to size plumbing and wiring that will provide the least power loss at the lowest cost.

The best way for a home hydro builder to design a system "is to solicit proposals from equipment suppliers," recommends Dan New. "Suppliers and manufacturers have been designing systems and handling problems for years, and that experience is available and free to the home developer."

Once the system is designed, suppliers can calculate how much electrical power the proposed system will generate.

BUYING AND INSTALLING A SYSTEM

If you are designing a system on your own, you'll need to consider each component of the system separately, starting with the intake structure. This section will help you understand a bit more about each component. If you are relying on a microhydro expert to design your system, he or she will specify the components and suggest the best design. Nonetheless, it is wise to read the material in this section so that you better understand what they're up to. The more informed you are, the more likely you are to get the system that you will be happy with over the long haul.

A Good Site

The key element for a good site is the vertical distance the water drops. A small amount of water dropping a large distance will produce as much energy as a large amount of water dropping a small distance. Thus, we can get approximately the same power output by running 1,000 gallons per minute through a 2-foot drop as by running 2 gallons per minute through a 1,000-foot drop.

— John Schaeffer, *Solar Living Source Book*

As you read the following material, remember that microhydro can be combined with any other renewable energy technology. For example, a system can combine PVs and microhydro, or a wind turbine and microhydro, or perhaps all three. Microhydro systems can even be coupled with grid power or a diesel or propane generator. Combining systems not only ensures better year-round service, it also may save you some money on one or more parts of your system. For example, combining PV and microhydro will probably mean that you'll need a smaller PV system and a smaller battery bank. With this in mind, let's look at the various components of a microhydro system, beginning with the intake structure.

Water Intake

All microhydro systems require an intake at the highest point in the system to divert water from the stream into the penstock. As noted earlier, all intakes need to be screened to prevent debris from entering the pipeline. Ideally, the screen should be self-cleansing. In other words, it should be designed so that debris washes off naturally. Otherwise, you'll have to manually remove debris blocking water flow into the pipeline — and do so regularly. Blockages can seriously reduce a system's output.

Screens are immersed in water 24 hours a day, so they should be made of a durable material that won't rust. One excellent choice is stainless steel. You can find stainless steel screens at agricultural supply outlets in rural areas where irrigation is common. If your agricultural supply outlet doesn't have screens, they may be able to order one for you. You may also want to check out commercially manufactured microhydro intake screens like those made by Hydroscreen (hydroscreen.com).

Intake structures vary in complexity. The simplest of all intakes is a screened pipe placed in a deep pool of water in the stream. The most complex intake systems consist of small concrete diversion structures or dams that block flow across an entire stream or river. The design, of course, depends on your site.

When designing an intake, it is important to remember the function of this structure. One important function of an intake structure is to prevent air from entering the pipeline. Air flowing through a turbine reduces power output and may damage the turbine. To be sure that no air enters the system, the intake pipe needs to be fully submersed all year long. This is done by placing the pipe in a deep pool, or building an intake structure that provides a pool that is deep enough all year round to prevent air from being sucked into the system — even during that part of the year when stream flows are at their lowest.

The intake structure also needs to provide a stilling (settling) basin — a pool that allows silt, sand, gravel, and rocks to settle out *before* stream water enters the penstock. Figure 10-4 shows a section through an intake structure. In this structure, the stilling basin empties into a second pool. Water flows from here into the pipeline. A screen over the pipe helps prevent debris from entering it.

Microhydro systems require very little maintenance. Sealed bearings in brushless generators and long-lasting runners mean that you won't be lubricating or replacing parts very often. You will, however, need to pay attention to the stilling basin and screen. Stilling basins may need to be emptied of debris, and screens may need to be cleaned regularly. To reduce routine cleaning, you should consider creating an overflow in the stilling pond that naturally siphons off floating debris and returns it to the stream. The screen should be as close to vertical as possible or slanted so that it self-cleans as much as possible. The intake screen and pipe also need to be submerged below the frost line to prevent freezing. Ice buildup on the screen could slow the flow of water into your system and sharply reduce its electrical output.

Choosing a Turbine/Generator

Several different types of microhydro turbine/generators are available. Your choice of turbine depends on the characteristics of your site. For mountainous sites that have fairly high pressure resulting from a favorable combination of head and water flow, you may want to consider the small Pelton wheel Harris systems (Figure 10-8). Pelton wheels are durable and require very infrequent replacement due to wear and tear. Even with dirty, gritty water, it takes at least ten years to wear out a Pelton wheel manufactured by Don Harris, according to Gaiam Real Goods. However, Harris offers a conventional automotive alternator with his system. These devices have bearings and brushes that need to be replaced at intervals of one to five years, depending on how hard the system works. To

Fig. 10-8:
Harris turbine.

minimize maintenance, Harris also offers a brushless permanent magnet generator. These units cost about $700 more, but use larger, more sturdy bearings that last two to three times longer. This greatly reduces maintenance and replacement costs. It is well worth the extra cost.

Fig. 10-9: *Low Head Stream Engine.*

Harris turbines are for applications with 50 feet or more of head. They produce from 1 kWh to 35 kWh per day,

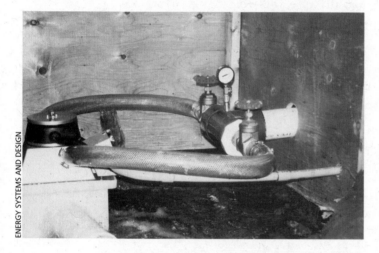

depending on the model and site. The unit (generator) has an instantaneous output of about 2,500 watts — enough to start most appliances and power tools. Harris microhydro turbines come in 24- and 48-volt models and are manufactured in the United States.

For low-head systems that channel a lot of water through a turbine with little head (that is, on flatter sites), you may want to consider a Low Head Stream Engine (Figure 10-9) or one of the newest microhydro turbine/generators, the Jack Rabbit (Figure 10-10). The Stream Engine is a brushless permanent magnet generator with sealed bearings that minimize maintenance. These units operate in sites with heads that range from 5 to 400 feet that have flow rates from 5 to 300 gallons per minute (although flow rates above 150 gpm provide very little additional power). Although set-up can be a bit tricky, the unit is virtually maintenance-free. Output voltages are 12, 24, and 48 volts. They're manufactured in Canada.

The Jack Rabbit is an unusual microhydro turbine and generator. As shown in Figure 10-10, this submarine-like device consists of a submersible propeller and generator in a waterproof casing that's suspended in a stream in at least 13 inches of water. The generator operates at low speeds, ideally around 9 miles per hour, or equivalent to a slow jog. Water flowing

Fig. 10-10: *Jack Rabbit.*

past the device causes the propeller to spin.

In slow-moving streams, water flows can be increased by piling rocks or heavy timbers to create a funnel with the wide end facing upstream. This channels water into the narrow end of the funnel, which greatly increases its speed. (This phenomenon is known as the *Venturi effect*.) The Jack Rabbit is placed at the narrow part of the funnel where the water is flowing most rapidly, optimizing the performance of the generator.

Jack Rabbits are low-capacity microhydro turbines that produce 100 watts of continuous power in a 9-mile-per-hour stream — or about 2.4 kWh of electricity per day — and slightly less in a slower stream. Jack Rabbits are rugged and require very little maintenance. The heavy-duty aluminum blades are not easily damaged. If struck by a log and bent, the propeller blades can be hammered back into shape. Another advantage of the Jack Rabbit is that it requires no pipeline, a feature that can save you a lot of money. Electrical lines from the generators (mounted securely in a stream) carry the current to the house. Jack Rabbits come in 12- and 24-volt models and are made in Great Britain. Proceed with caution with these, however. Word from users is that the power output is very low, and there are other problems that may make them a dubious choice.

Finally, the Low Head Stream Engine, with a turbo-type runner, is for use in sites that offer intermediate characteristics. They operate in areas with drops of 2 to 10 feet and flows ranging from 200 to 1,000 gallons per minute. Peak output is 1,000 watts, meaning that a turbine operating in a good site will provide 24 kWh of electricity per day.

To learn more about these systems, I strongly recommend that you consult John Schaeffer's book, *Solar Living Source Book*. It provides a good overview of the subject and a lot of valuable information on the various microhydro turbines, including pricing. Remember, as Dan New points out, that "the turbine should be designed to match your specific head and flow. Proper selection requires considerable expertise."

AC Generators

Although most microhydro systems produce DC electricity, there are some times when AC generators are advisable, notably for larger commercial applications or for grid-connected systems. These systems can produce more than 3,000 watts of continuous power or 72 kWh per day — far more than most homes ever consume. As Dan New pointed out in an e-mail to me, these systems "can often supply all of the energy needs of a home." DC systems typically only provide electricity for lighting, appliances, and electronic heaters, but not space heat, hot water, or dryers. Those demands are typically satisfied by propane or natural gas. An AC microhydro system, however, could supply heavy-duty machines, compressors, and welding and woodworking equipment. "Most DC systems cannot cover most of this list," concludes New, so homeowners must rely "on nonrenewable sources of energy to accomplish these tasks."

According to Schaeffer, AC systems often involve considerable engineering, custom metalwork, formed concrete, permits, and a fairly high initial investment. AC systems also require governors to ensure that the AC generator spins at the correct speed to produce electricity of the proper frequency to match household needs. In grid-connected systems, additional controls are required to ensure that power going onto the electric grid matches line current with respect to voltage, frequency, and phase. These controls will automatically disconnect the system from the grid if major fluctuations occur on either end. Although AC systems cost more upfront, "AC power is almost always cheaper per kilowatt-hour," says New. AC hydro systems also require no batteries and do not require inverters. AC power is used directly as it is produced. Surplus can be diverted to dump loads, usually resistance heaters, or can be diverted to the electrical grid. "So," says New, "if a homeowner can find a use for most of the power most of the time, a moderately sized AC system, say 4 kW to 15 kW, will provide the energy needed to power a home renewably for less money." (For details on grid-connected renewable energy systems, see Chapter 8.)

For those who want battery backup, in case the system must be shut down, remember that AC power can be converted to DC power by a bank of diodes and then stored in a battery bank for later use. However, also remember that to be useful, DC electricity from the battery bank must be converted back to AC power. That's a function of the inverter. As Dan New pointed out to me, systems that convert AC to DC power are rare.

AC systems can be quite costly, warns Schaeffer, so this "isn't an undertaking for the faint-of-heart or thin-of-wallet." This is not to say these systems are uneconomical. Quite the opposite: they often provide a very good return on investment. "Far more streams can support a small DC-output system," says New, "than a higher output AC system." Those who think they have a site that is suitable for low-head AC power production, can download information on these systems from the following website: eere.energy.gov/RE/hydropower.html. For more on the cost issue, see the sidebar, "AC vs DC Costs."

Powerhouse

As noted earlier, turbines and generators must be protected from the elements to ensure long service. Protection is provided by a powerhouse. Powerhouses range in complexity from simple wooden or cement block boxes (designed so you can work on the system) to small sheds with room for battery banks.

Be sure when connecting pipes to the turbine in a powerhouse to use as few elbows as possible. Bends in pipe dramatically reduce head by creating turbulence. Reduced head, in turn, reduces system output. "Likewise," notes Dan New, "any restrictions on water exiting the turbine may increase resistance against the turbine's moving parts." This, too, reduces power output.

Drive Systems

In most systems, a steel shaft connects the turbine to the generator. This shaft couples the turbine with the generator so that rotation of the turbine's runner translates into rotation within the generator. "The most efficient and reliable drive system involves a direct 1:1 coupling between the turbine and generator," notes New. But this is not possible for all sites. In some cases, especially when AC generators are used, it may be necessary to "adjust the transfer ratio so that both the turbine and generator run at their optimum (but different)

AC vs DC Costs

Although an AC microhydro system typically costs more than a DC system, this comparison isn't fair. You're not comparing apples to apples. Most DC systems don't produce anywhere near as much electricity as AC systems. The result is that you will very likely need to acquire additional energy to run your home, for example, natural gas. You may also need to install additional equipment, for example, a wood stove to provide heat or a gas or diesel generator to provide additional electricity to run a washing machine.

speeds," he adds. This is achieved through gears, chains, or belts. Belt systems are the most popular because they are the least expensive. Unfortunately, more complex drive systems increase the cost of a system and will invariably increase maintenance requirements.

Batteries

For those who need lots of power intermittently (which is most of us) a battery system or a grid-connected system may be advisable.

Most microhydro systems use deep-cycle lead acid batteries. Never, never, never use automobile batteries. They can't handle the deep discharging. Batteries for renewable energy systems were covered at length in Chapter 8, but I'll point out one important fact here. As a general rule, microhydro systems require much smaller, and thus much less expensive, battery banks than solar electric or wind electric systems. That's because these systems only need to provide electricity for occasional heavy power usages and power surges. You're not trying to store power for three to four days of cloudy weather, as in the case of a PV system, or windless days, in the case of a wind energy system. The battery bank is also smaller because of rapid recharge. That is, if the batteries are drawn down during the day, they're usually recharged by evening. Occasional high output and rapid recharge not only means fewer batteries, it also means a longer battery life, for reasons explained in Chapter 8.

Pipeline

After designing an intake structure and selecting a turbine, you'll need to design a pipeline. Pipelines are typically made from either 4-inch PVC or smaller polyethylene pipe. PVC is used almost exclusively when the pipe needs to be over 2 inches in diameter, although 1.5- to 2-inch PVC can be and is used. Four-inch PVC pipe comes in 10- and 20-foot sections that are glued together. Assembly is quick and painless and can be mastered by anyone. PVC pipe not only goes together easily, it is relatively inexpensive. In addition, PVC pipe is very light, so it is easy to install, which is especially helpful in steep terrain. Two-inch polyethylene pipe is used for smaller flows. It comes in very long rolls that are laid out from intake to turbine. Because there's no gluing (unless two rolls must be connected), polyethylene pipe goes in much faster than PVC.

Although plastic pipe is fairly inexpensive, the pipeline can be a costly and time-consuming aspect of a microhydro system. "It's not unusual to use several thousand feet of pipe to collect a hundred feet of head," notes Schaeffer. Additionally, in cold climates, pipe may need to be buried below the frost line to prevent it from freezing. This can add significantly to the cost and labor required to install a system, especially if the soils are rocky and difficult to work in, which is often the case in mountainous terrain. However, Scott Davis points out that "even if the pipe is not quite below the frost line, water running through the pipe may keep it from freezing."

Burying the pipeline not only protects the pipe from freezing, it helps protect it from damage, for instance, from falling trees or tree limbs. And it helps keep the pipe from shifting around as high-pressure water flows through it. PVC pipe deteriorates in sunlight, too, necessitating burial.

Controllers

Microhydro systems require special controllers to prevent batteries from overcharging, and hence from being permanently damaged. Unlike the charge controllers on PV systems that terminate the flow of electricity to the batteries, microhydro system controllers shunt excess electricity to a secondary load, typically an electric resistor or two (Figure 10-11). These resistors are typically water- or space-heating elements that put excess electricity to good use heating domestic hot water or the house. They are referred to as *dump loads.* (Note: PV charge controllers should not be used in a microhydro system, as they could damage the generator.)

Excess power is shunted to secondary loads in off-grid systems when the system's batteries, if any, are full and when the household demands are being met by the system. In grid-connected systems, excess power is diverted onto the grid.

Load controllers and water heater and air heater diversion devices can be purchased from Gaiam Real Goods and other suppliers that cater to folks interested in microhydro.

Power Lines

Most microhydro systems produce low-voltage DC electricity that travels in a wire to the battery bank. Because low-voltage current doesn't travel well — it loses energy quickly over distance — it is best to keep power lines from the turbine/generator short, under 100 feet. If this is not possible, you will need to use large-gauge wire, which costs much more than standard wire.

CAUTION ''' HOT '''

AEE SOLAR

Fig. 10-11: *Load diverters (dump loads) take excess energy and run it through resistors (shown on the right) to generate heat. They're useful in the winter to provide heat to living spaces as well as garages, sheds, and workshops.*

Permitting

Jeffe Aronson installed a low-head microhydro system on a stream running through his property in Australia. He did so by building a small concrete dam across part of the river. Although friends recommended that he complete the project without permits and "let the bureaucrats find it if they could," Jeffe decided to secure the necessary permits and work with local authorities. In fact, he made a point of working with them in a cooperative fashion — and not showing anger or frustration at some of their quirks. In short, he established a good, respectful working relationship. "A couple years later," Aronson writes in *Home Power,* Issue 101, a visiting angler, "who had fished this section of river for decades and considered it his own, came upon our works." The angler was outraged and, rather than consult Jeffe, he sent a complaint to the water catchment authority. "They sent a representative with whom we'd dealt originally," said Aronson. "He thankfully found that we'd done what we'd said we'd do and even felt the works to be 'very discreet.'" And that was the end of the story. The lesson in this tale is that permitting may be a bother, but it can save you a lot of hassle and expense. In fact, government agencies can force a landowner to remove a system that cost several thousand dollars to install, if they catch him or her generating microhydro without a permit.

Ask a local supplier or installer to find out what permits are required. If there are no local experts, you will need to call the state's office of environmental protection or state energy office. They can steer you in the right direction.

FINDING AN INSTALLER OR INSTALLING A SYSTEM YOURSELF

Microhydro systems are pretty easy to install. In fact, people with modest plumbing and electrical skills can install most of a system without much trouble. However, battery banks, inverters, and other controls typically used in off-grid systems require a higher degree of expertise than many do-it-yourselfers possess. Grid-connected systems require a high level of electrical expertise too. My advice is that it's always a

good idea to hire a professional to help you out or to do the job for you. Experienced professionals know the tricks of the trade and can save you a lot of time, trouble, expense, and frustration. I've always been amazed when working with experts on wind and solar systems, for instance, how little tricks of the trade make the job go so much easier!

A professional installation will cost more, but the added expense could very likely be well worth it. Not only will you probably get a better system, you'll have someone to call in case something goes wrong and you can't fix it.

To locate a professional, contact a local renewable energy society or, if there is none, call a manufacturer or Real Goods and ask for references in your area.

THE PROS AND CONS OF MICROHYDRO SYSTEMS

Microhydro is a great source of reliable electrical energy, but before you go out to buy a system, you should have a full understanding of its pros and cons. On the plus side, microhydro is probably *the* most cost-effective renewable energy system on the market. According to Scott Davis, it delivers "the best bang per buck." Davis goes on to say, "Significant power can be generated with flows of two gallons per minute or from drops as small as two feet." And, says Davis, "power can be delivered in a cost-effective fashion a mile or more from where it is generated to where it is being used."

Microhydro, unlike almost all other renewable forms of energy (certainly wind and solar) often provides continuous power. That is to say, it provides electricity day and night, night and day, 7 days a week, 365 days a year. Excess power can be stored in batteries or can be fed onto the utility grid, if it's nearby. As Figure 10-12 shows, microhydro can do a much better job of meeting one's needs. Solar and wind systems require either considerable

Fig. 10-12:
Comparison of microhydro and hybrid solar/wind systems.

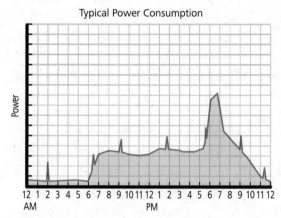

Typical Power Consumption

Your electrical needs vary over time.

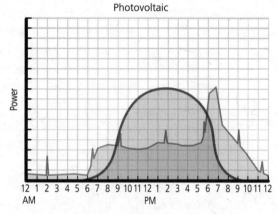

Photovoltaic

PV outputs drop to zero every night and are much reduced in winter and cloudy weather.

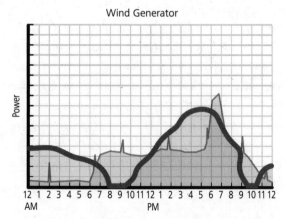

Wind Generator

Wind system outputs go up and down irregularly.

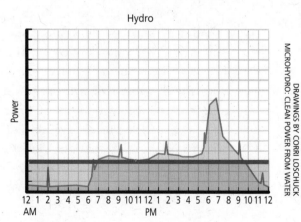

Hydro

Microhydro systems have the same output all the time.

DRAWINGS BY CORRI LOSCHUCK
MICROHYDRO: CLEAN POWER FROM WATER

storage capacity or a connection to the electrical grid to supply continuous power. Microhydro sites owe their year-round performance in part to the fact that they can be easily winterized. Even though stream flows may slow down during the winter in snowy climates, small streams continue to offer reliable water flows because moving water doesn't freeze very readily. Burying pipes underground below the frost line also helps to ensure a steady flow. Furthermore, notes Davis, microhydro can be integrated with pipelines delivering water to a house from a spring or stream located uphill from the building. small microhydro systems are inexpensive, too, costing from $2,000 to $10,000.

Yet another advantage of microhydro is that it is applicable to a variety of different conditions — from high head/low flow to low head/high flow and everything in between. The rule of thumb is, the more head, the less volume will be necessary to produce a given amount of power. The less head, the more volume you'll need.

Like all systems, there are some disadvantages to microhydro systems. First off, very few suitable sites are available, say, in comparison to solar electric or wind energy sites. Second, microhydro systems require considerable knowledge. Third, not all sites that could be used for microhydro can be developed economically. Pipelines for a system, for instance, may be long and thus costly. Or they may be difficult to install. Fourth, diverting water from streams can alter flows and adversely affect living organisms that rely on the water. You need to exert extreme caution when installing a microhydro system. Be sure to consult with a qualified stream biologist. Fifth, microhydro systems may require permits from local government agencies, which can take time and may require a financial outlay on your part — notably, for engineering costs or the costs of a biologist to examine the site for potential impact. In British Columbia, one of my clients had two large streams flowing on either side of his property but could not tap into the flowing water because the province prohibits structures on streams to protect fish populations.

As with any system, you need to proceed carefully and cautiously — with as complete an understanding of the system as possible. If you are lucky enough to live by a stream or river and can legally tap into the power of the water running by your home, you could be graced with years of very inexpensive, clean, and renewable energy.

A Brief Afterword

In this book, we've explored many ways to conserve energy and trim demand by using energy more efficiently in your home or business to get the most for your energy dollar and also to help create a cleaner, healthier, and more sustainable future. We've also explored a wide range of options for tapping into renewable energy. By now, you should have a pretty good idea what ideas you want to pursue. If you are going to pursue some of these projects on your own, you may want to sign up for a few hands-on workshops first. It will make your life much easier in the long run.

Bear in mind that home energy efficiency and residential renewable energy are just a few of the ways to achieve greater self-reliance. There's much more you can do to achieve a greater level of independence, like growing some of your own food, driving a greener vehicle or using a green fuel like straight vegetable oil or ethanol, but those are topics of other books. For ideas on transportation, you may want to check out my most recent book, *Green Transportation Basics*. It is full of useful information and advice on sustainable fuels and sustainable vehicles.

For ideas on greening your home and your life, I recommend that you take a look at a book I co-authored with Dave Wann, called *Superbia! 31 Ways to Create Sustainable Neighborhoods*. It offers a wealth of advice on many other aspects of your life — and ways you can get your whole neighborhood involved. And my book *Green Home Improvement* contains

65 projects you can do to green up your home.

At this point, it is up to you. Study your options carefully. Keep your eye on the prize always, and proceed forward with the confidence that you're not only helping to create a better future for yourself and your family, you're helping to build a sustainable future for all human beings and the millions of species that share this planet with us.

This is an enormously important task. Although it may seem like you are only one of a very few who care or who have made a commitment to take action, I think you would be surprised to learn how many fellow citizens share your passions. You can be a model to them — something of a catalyst that spurs them into action. As always,

I wish you the best of luck!

Appendix

Metric Conversions		
If you know:	**Multiply by:**	**To find:**
Inches	2.5	centimeters
Feet	0.3	meters
Miles	1.6	kilometers
Gallons	4.5	liters
Pounds	0.45	kilograms
Acres	0.4	hectares
Square feet	0.09	square meters
Miles per hour	1.6	kilometers per hour

To convert miles per gallon to liters per 100 kilometers, divide 282 by the number of miles per gallon.
For example, 30 miles per gallon is 282/30 = 9.4 liters per 100 kilometers.

Resource Guide

This resource guide contains a wealth of information on renewable energy listed by chapter. Information is periodically updated on my website: danchiras.com.

Introduction

Publications

Campbell, C. J. *The Coming Oil Crisis*, Petroconsultants Sa, 1997. Detailed and authoritative analysis of world oil supplies. Paints a rather dismal future.

Campbell, C. J. "The Oil Peak: A Turning Point," *Solar Today* 15(4), 2001. A brief analysis of the future of global oil.

Darley, Julian. *High Noon for Natural Gas: The New Energy Crisis*. Chelsea Green, 2004. A look at natural gas supplies and the impacts of the imminent peak in global natural gas production.

Hartmann, Thom. *The Last Hours of Ancient Sunlight: Warming Up to Personal and Global Transformation*. Three Rivers Press, 1999. A philosophic look at the decline in oil.

Heinberg, Richard. *The Party's Over: Oil, War and the Fate of Industrial Societies*. (revised and updated version) New Society Publishers, 2005. A grim look at the prospects for human society as global oil production peaks.

Heinberg, Richard. *Power Down: Options and Actions for a Post-Carbon World*. New Society Publishers, 2004. A frank discussion of our options as global oil production peaks.

Howe, John. *The End of Fossil Energy: A Plan for Sustainability*. McIntire Publishing, 2003. A short overview of energy problems and what we can do about them.

Rifkin, Jeremy. *The Hydrogen Economy: The Creation of the World-Wide Energy Web and the Redistribution of Power on Earth.* Tarcher/Putnam, 2002. Excellent analysis of world energy troubles and an optimistic look at the role hydrogen could play in our future.

Udall, Randy. "From Cleopatra to Columbia," *Home Power* 100, 2004. An insightful look at energy in an historical perspective.

Udall, Randy and Steve Andrews. "Methane Madness: A Natural Gas Primer," *Solar Today* 15(4), 2001. A frightening look at natural gas production.

Udall, Randy and Steve Andrews. "When Will the Joyride End?" *Home Power* 81, 2001. An insightful look at oil production and consumption with emphasis on peak oil production.

Videos

The End of Suburbia: Oil Depletion and the Collapse of The American Dream. A gripping video on the impacts of peak oil and natural gas production. Paints a dim future for human culture. To order a copy contact: endofsurburbia.com.

Organizations

Global Public Media. This organization helps to spread the word about peak oil and natural gas and tracks important developments. Website: globalpublicmedia.com. PeakOilAction.org is another group dedicated to helping spread the word about the proximate peaks in global oil and natural gas production. Website: peakoilaction.org.

The Post Carbon Institute is working to spread the word about the proximate peaks in global oil and natural gas production. Website: postcarbon.org.

Chapter 1: Renewable Energy — Clean, Affordable, and Reliable
Publications

Aitken, Donald W. "Germany Launches Its Transition," *Solar Today* 19(2), 2005. A fascinating look at Germany's plans to achieve 100 percent of its energy from renewable sources.

Aitken, Donald W. "The Renewable Energy Transition: Can it Really Happen?" *Solar Today* 19(1), 2005. A valuable look at a very important question.

Asmus, Peter. "Power Solutions in Our Own Backyards," *Solar Today* 15 (1), 2001. An interesting look at dispersed energy production.

Bihn, Dan. "Japan Takes the Lead," *Solar Today* 19(1), 2005. A look at Japan's aggressive pursuit of renewable energy.

Brown, Lester R. "Harnessing the Wind for Energy Independence," *Solar Today* 16(2), 2002. A good overview of wind energy.

Brown, Lester R. "The Short Path to Oil Independence," *Mother Earth News* 208, 2005. How hybrid cars and wind energy can slash our oil imports.

Chiras, Daniel D. *Lessons From Nature: Learning to Live Sustainably on the Earth.* Island Press, 1992. Describes transitional steps to a renewable energy economy.

Chiras, Dan. "All About Insulation," *Mother Earth News* 195, 2003. A detailed look at insulation options.

Chiras, Dan and John Richter. "Money Matters: Does an RE System Make Economic Sense?" *Home Power* 131, 82–85, 2009.

Flavin, C. "Rethinking the Energy System," In *State of the World 1999.* Linda Starke, ed., Norton, 1999. Part of an ongoing series of books on ways to restructure the global energy system that are good for people and the environment.

Home Power Staff. "Clearing the Air: Home Power Dispels the Top RE Myths," *Home Power* 100, 2004. Superb piece for those who want to learn more about the many false assumptions and beliefs about renewable energy.

Kohn, Lin. "Solar in San Francisco," *Solar Today* 1 (4), 2003. An exciting look at what one progressive city is doing to increase its reliance on renewable energy.

Perlin, J. *From Space to Earth: The Story of Solar Electricity.* Aatec Publications, 1999. A wonderfully readable history of solar electricity.

Perlin, John, Lawrence Kazmerski, and Susan Moon. "Good as Gold: The Silicon Solar Cell Turns 50," *Solar Today* 28(1), 2004. A brief history of PV research and development.

Prugh, Tom, Christopher Flavin, and Janet L. Sawin. "Changing the Oil Economy," In *State of the World 2005.* Linda Starke, ed., Norton, 2005. Examines oil production declines from the standpoint of national security.

Renner, M. "Creating Jobs, Preserving the Environment," In *State of the World 2000.* Linda Starke, ed., Norton, 2000. Shows the economic and employment benefits of renewable energy.

Renner, M. "Employment in Wind Power." *World Watch* 14(1), 2001. Fascinating look at the employment boom in the wind energy sector.

Roodman, D. M. "Reforming Subsidies." In *State of the World 2000.* Linda Starke, ed., Norton, 2000. Extremely important reading.

Sawin, Janet L. "Charting a New Energy Future," In *State of the World 2003.* Linda Starke, ed., Norton, 2003. An overview of renewable energy potential and what other nations are doing to create a sustainable energy future.

Sawin, Janet L. *Mainstreaming Renewable Energy in the 21st Century*, Worldwatch Paper 169, Washington, D.C.: Worldwatch Institute, 2004. Overview of the prospects of renewable energy.

Sawin, Janet L. "Making Better Energy Choices," In *State of the World 2004*. Linda Starke, ed., Norton, 2004. An overview of what people are doing worldwide to switch to a renewable energy strategy.

Sindelar, Allan and Phil Campbell-Graves. "How to Finance Your Renewable Energy Home," *Home Power* 103, 2004. Very useful article.

Starrs, Tom. "Green Tags: A New Way to Support Renewable Energy," *Solar Today* 15 (4), 2001. Describes ways that individuals can support renewable energy production without having to install a system on their homes.

Stronberg, Joel B. "A Common-Sense Solution," *Solar Today* 19(2), 2005. A look at the renewable energy strategy in light of current affairs.

Strong, Steven J. "Beyond Petroleum," *Solar Today* 15(2), 2001. One of my all-time favorite stories: British Petroleum's shift to renewable energy.

Swain, J. "Charting a New Energy Future," In *State of the World 2003*. Linda Starke, ed., Norton, 2003. Examines the potential of renewable energy resources and many other important subjects.

Swezey, Blair and Lori Bird. "Businesses Lead the 'Green Power' Challenge," *Solar Today* 15(1), 2001. What some major corporations like Toyota are doing to increase their reliance on renewable energy.

Swezey, Blair and Lori Bird. "Buying Green Power — You Really Can Make a Difference," *Solar Today* 17(1), 2003. An in-depth look at ways, such as green tags, that homeowners can tap into renewable energy without installing a system on their home.

Wilkerson, Mark W. "Solar in the Heartland," *Solar Today* 14(5), 2000. An inspiring tale of a small unincorporated village in north central Illinois that has gone solar.

Yerkes, Bill. "40 Years of Solar Power," *Solar Today* 18(1), 2004. A historical look at PV development from an insider.

Yewdall, Zeke. "Alternatives that Don't Cost an Arm and a Leg," *Home Power* 101, 2004. A useful article on energy efficiency and other ways to acquire electricity from renewable sources if your site is not ideally suited for them or your pocketbook prohibits the investment.

Chapter 2: Conservation Rules — The Cornerstone of Your Energy Future

Publications

Bailes, Allison A. "How Efficient is Your House?" *Home Power* 106, 2005.

Great article on home energy ratings and blower door tests.

Barker, Jennifer. "Buying an Energy-Efficient Refrigerator," *Home Power* 104, 2005. For those interested in upgrading their refrigerators, this article is a must.

Carmody, John, Stephen Selkowitz, and Lisa Heschong. *Residential Windows: A Guide to New Technologies and Energy Performance*. Norton, 1996. Important resource for energy-efficient home designers.

Chiras, Dan. "Hunting Phantom Loads," *Home Power* 82, 2001. My own experience with phantom loads in my solar electric home and advice on eliminating them.

Chiras, Dan. "Minimize the Digging: Frost- Protected Shallow Foundations," *The Last Straw* 38, 2002. A brief overview of frost-protected shallow foundations.

Chiras, Dan. "Retrofitting a Foundation for Energy Efficiency, *The Last Straw* 38, 2002. Describes ways to retrofit foundations to reduce heat loss.

Chiras, Daniel D. "The Energy-Efficient Home," *Solar Today* 18(5), 2004. A primer on energy-efficient home design.

Connor, Rachel. "Basics to Building a Better Home," *Home Power* 126, 40–45, 2008. Extremely important reading to those interested in minimizing energy consumption in a home or business.

Fine Homebuilding. *The Best of Fine Homebuilding: Energy-Efficient Building*. Taunton Press, 1999. A collection of detailed articles on a wide assortment of topics related to energy efficiency including insulation, energy-saving details, windows, and heating systems.

Hurst-Wajszczuk, Joe. "Save Energy and Money — Now!" *Mother Earth News* 189, 2001. More ideas on ways to save energy in your home.

Kerr, Andy. "Doing Well While Doing Good: Conservation of Energy as a Rational Financial Investment," *Home Power* 86, 2002. A look at the economics of household energy savings.

Johnston, David and Kim Master. *Green Remodeling: Changing the World One Room at a Time*. New Society Publishers, 2004. Superb coverage of many ideas on ways to boost energy efficiency in existing homes.

Johnston, David and Scott Gibson. "Green from the Ground Up," *Home Power* 129, 66–70, 2009. This article contains a lot of useful information on foundations for energy-efficient homes.

Johnston, David and Scott Gibson. "Green Framing Options," *Home Power* 130, 58–64, 2009. Important information on ways to build more energy efficient homes.

Lstiburek, Joe, and Betsy Pettit. *EEBA Builder's Guide — Cold Climate*. Energy Efficient Building Association, 1999.

Superb resource for advice on building in cold climates.

Lstiburek, Joe, and Betsy Pettit. *EEBA Builder's Guide — Mixed Humid Climate*. Energy Efficient Building Association, 1999. Superb resource for advice on this climate.

Lstiburek, Joe, and Betsy Pettit. *EEBA Builder's Guide — Hot-Arid Climate*. EEBA, 1999. Superb resource for advice on building in hot, arid climates.

Magwood, Chris, ed., "Roofs and Foundations," *The Last Straw* 38, 2002. An excellent resource for those who want to learn about efficient foundations.

National Association of Home Builders Research Center. *Design Guide for Frost-Protected Shallow Foundations*. NAHB Research Center, 1996. Also available online at nahb.org.

Pahl, Greg. *Natural Home Heating: The Complete Guide to Renewable Energy Options*. Chelsea Green, 2003. A detailed overview of natural home heating options.

Pinkham, Linda and Joe Schwartz. "Washing Machine Spin-Off: Maytag Neptune vs. Frigidaire Gallery vs. Thrift Store Model," *Home Power* 103, 2004. A good comparison of several of your options for energy and water-efficient washing machines.

Salomon, Thierry and Stephane Bedel. *The Energy Saving House*. Centre for Alternative Technology Publications, 1999. Good guide to home energy savings.

Scheckel, Paul. *The Home Energy Diet*. New Society Publishers, 2005. A great guide for energy conservation in homes.

Scheckel, Paul. "Efficiency Details for a Clean Energy Change," *Home Power* 121, 40–45, 2007. Ten fantastic tips on saving energy.

Schwartz, Joe and Doug Puffer. "The Perfect PV: Home Power's 2007 Solar-Electric Module Guide," *Home Power* 121, 70–78, 2007. Check this out when buying PV modules. Look for the updated version published every year in *Home Power*.

Sikora, Jeannie L. *Profit from Building Green: Award Winning Tips to Build Energy Efficient Homes*. BuilderBooks, 2002. A brief, but informative overview of energy conservation strategies.

Williams, Jeff. "Warm Light, Cool Savings," *Solar Today* 19(2), 2005. An introduction to energy-efficient windows and doors.

Wilson, Alex. "Windows: Looking through the Options," *Solar Today* 15(2), 2001. A great overview of windows with a useful checklist for those in the market to buy new windows.

Organizations

American Council for an Energy-Efficient Economy. 1001 Connecticut Avenue

NW, Suite 801, Washington, D.C. 20036. Tel: (202) 429-0063, Website: aceee.org. Numerous excellent publications on energy efficiency, including *Consumer Guide to Home Energy Savings.*

Building America Program. US Department of Energy. Office of Building Systems, EE-41, 1000 Independence Avenue SW, Washington, D.C. 20585. Tel: (202) 586-9472. Leaders in promoting energy efficiency and renewable energy to achieve zero-energy buildings.

Cellulose Insulation Manufacturers Association. Your place to shop for information on cellulose insulation. 133 S. Keowee St., Dayton, OH 45402. Tel: (937) 222-2462. Website: cellulose.org.

Consumers Union. Tel: (800) 500-9760. Publishes *Consumer Reports* and *Consumer Reports Annual Buying Guide,* which rates appliances for reliability, convenience, and efficiency. Website: consumerreports.org.

Energy Efficiency and Renewable Energy Clearinghouse. P.O. Box 3048, Merrifield, VA 22116. Tel: (800) 363-3732. Great source for a variety of useful information on energy efficiency.

Energy Efficient Building Association. 490 Concordia Ave., P.O. Box 22307, Eagen, MN 55122. Tel: (651) 268-7585. Offers conferences, workshops, publications and an online bookstore. Website: eeba.org.

National Fenestration Ratings Council. 1300 Spring St., Suite 120, Silver Springs, Md. Tel: (301) 589-6372. For information on the energy efficiency of windows. Website: nfrc.org.

National Insulation Association. 99 Canal Center Plaza, Suite 222, Alexandria, VA 22314. Tel: (703) 683-6422. Offers a wide range of information on different types of insulation. Website: insulation.org.

US Department of Energy and Environmental Protection Agency's ENERGY STAR program. Tel: (888) 782-7937. Website: energystar.gov. Energy-efficient appliances and heating systems

Publications

Fine Homebuilding. *Energy-Efficient Building.* Taunton Press, 1999. Contains a collection of extremely useful articles on mechanical heating systems.

Fust, Art. "A Simple Warm Floor Heating System," *The Last Straw* 32, 2000. Contains much useful information on radiant floor heat.

Grahl, Christine L. "The Radiant Flooring Revolution," *Environmental Design and Construction* January/February, 2000. Superb introduction to radiant-floor heating.

Hyatt, Rod. "Hydronic Heating on Renewable Energy," *Home Power* 79,

2000. Practical advice on installing a radiant-floor heating system using renewable energy.

O'Connell, John, and Bruce Harley. "Choosing Ductwork," *Fine Homebuilding* June/July, 1997. Essential reading for anyone interested in installing a forced-air heating system.

"Hydronic Radiant-Floor Heating," *Fine Homebuilding*, October/November, 1996. Extremely useful reference. Well-written, thorough, and well-illustrated.

Siegenthaler, John. *Modern Hydronic Heating*. Delmar Publishers, 1995. Everything you would ever want to know about hydronic heating.

Thorne, Jennifer, John Morrill, and Alex Wilson. *Consumer Guide to Home Energy Savings*, 8*th* ed. American Council for an Energy-Efficient Economy, 2003. Excellent resource, full of vital information on energy-saving appliances.

Weaver, Jennifer. "Tankless Is In," *Home Power* 105, 2005. Excellent overview of tankless water heaters.

Wilson, Alex. "A Primer on Heating Systems," *Fine Homebuilding*, February/March, 1997. Superb overview of furnaces, boilers, and heating systems.

Wilson, Alex. "Radiant-Floor Heating: When It Does — and Doesn't — Make Sense," *Environmental Building News* 1, 2002. Valuable reading.

Organizations

Consumer Product Safety Commission. Office of Information and Public Affairs, CPSC, Washington, D.C. 20207 or call their hotline at (800) 638-2772. Offers a wealth of information on space heaters, including safety precautions. Website: cpsc.gov.

Radiant Panel Association. 1433 West 29*th* Street, Loveland, CO 80539. Tel: (970) 613-0100. Professional organization consisting of radiant heating and cooling contractors, wholesalers, manufacturers, and professionals. Website: radiantpanelassociation.org.

Energy-Efficient Landscaping
Publications

Dramstad, Wenche E., James D. Olson, and Richard T. Forman. *Landscape Ecology: Principles in Landscape Architecture and Land- Use Planning*. Island Press, 1996. A useful textbook on the subject.

Moffat, Anne Simon, Marc Schiler and the staff of Green Living. *Energy-Efficient and Environmental Landscaping*. Appropriate Solutions Press, 1994. Contains an abundance of information on energy-efficient landscaping strategies and plant varieties suitable for various climate zones.

NREL. *Landscaping for Energy Efficiency*. DOE Office of Energy Efficiency and Renewable Energy, 1995. DOE/

GO-10095-046. Provides a decent, though somewhat disorganized, overview on the topic.

Striefel, Jan and Wesley A. Groesbeck. *The Resource Guide to Sustainable Landscapes.* Environmental Resources, Inc., 1995. Excellent resource.

Magazines

Earthwood Journal. Eos Institute, 580 Broadway, Suite 200, Laguna Beach, Ca. 92651. Phone: 1 (714) 497-1896. A glossy permaculture magazine published by Eos Institute and the Permaculture Institute of Southern California. Geared to the professional designer, architect, and land-use planner.

The Permaculture Activist. P.O. Box 1209W, Black Mountain, NC, 28711. Tel: (828) 298-2812. Publishes articles on a variety of subjects related to permaculture; includes an updated list of permaculture design courses. Website: permacultureactivist.net.

Permaculture Drylands Journal. Permaculture Drylands Institute, P.O. Box 156, Santa Fe, NM 87504. Tel: (505) 983-0663. Quarterly journal that focuses on the practice of permaculture in arid lands, especially Arizona and New Mexico. Website: permaculture.net.

Permaculture Edge. Permaculture Nambour, Inc. P.O. Box 148, Inglewood 6050, Western Australia. Reports on cutting-edge developments in permaculture.

Permaculture Magazine. Permanent Publications, Hyden House Limited, Little Hyden Land, Clandfield, Hampshire PO8 0RU, England. A quarterly journal published in cooperation with the Permaculture Association of Great Britain, containing articles, book reviews, and solutions from Britain and Europe. Website: permaculture.co.uk.

Permaculture Resources. P.O. Box 65, 56 Farmersville Road, Califon, NJ, 07830. Tel: (800) 832-6285. An educational publisher and distributor of permaculture resources and publications.

The Last Straw. TLS, P.O. Box 22706, Lincoln, NE 68542. Tel: (402) 483-5135. This journal publishes articles on natural building and features articles on passive solar heating and cooling. Website: strawhomes.com.

Organizations

Appropriate Technology Transfer for Rural Areas, P.O. Box 3657, Fayetteville, AR 72702. Tel: (800) 346-9140. This organization is actively involved in the permaculture movement.

International Permaculture Institute. An international coordinating organization for permaculture activities such as accreditation. P.O. Box 1, Tyalgum, NSW 2484, Australia. Tel: (066) 793 442.

Chapter 3: Solar Hot Water Systems — Satisfying Domestic Hot Water Needs with Solar Energy

Publications

Butler, Barry. "Solar Wand: Hot Water Assist for Cold Climates," *Home Power* 104, 2005. Illustrates and describes a quick retrofit for existing water tanks.

Galloway, Terry. *Solar House: A Guide for the Solar Designer*. Architectural Press, 2004. A technical guide on a wide range of solar designs, including active solar.

Guevara-Stone, Laurie and Ian Woofenden. "Choosing Your RE Installer," *Home Power* 127, 48–52, 2008. Read this article before you start calling potential installers.

Lane, Tom. "Solar Pool Heating Basics, Part 1," *Home Power* 94, 2003. Excellent overview of solar pool heating systems.

Lane, Tom. "Solar Pool Heating Basics, Part 2," *Home Power* 95, 2003. Excellent overview of solar pool heating systems.

Johnston, David and Kim Master. *Green Remodeling: Changing the World One Room at a Time*. New Society Publishers, 2004. Contains many ideas to boost energy efficiency in existing homes and increase one's reliance on renewable energy, including active solar.

Lane, Tom and Ken Olson. "Solar Hot Water for Cold Climates, Part II: Drainback Systems," *Home Power* 86, 2002. Detailed look at drainback systems. (Part I of this series is Olson's article on closed-loop antifreeze systems, listed below. Part III, written by Marken and Olson, is also listed below.)

Marken, Chuck. "Heat Exchangers for Solar Water Heating," *Home Power* 92, 2003. Great overview of heat exchangers.

Marken, Chuck. "New Life for Your Old Water Heater: Water Heater and Solar Tank Anode Rods," *Home Power* 106, 2005. A must read for anyone who has a conventional water heater.

Marken, Chuck. "Get Into Hot Water: Home Power's 2008 Solar Thermal Collector Guide," *Home Power* 123, 66–74, 2008. Just like other *Home Power* consumer guides, this one is priceless. Be sure to look for periodic updates of this guide in more recent issues of the magazine.

Marken, Chuck. "Solar Collectors: Behind the Glass," *Home Power* 133, 70–76, 2009. A detailed look at solar hot water collectors.

Marken, Chuck and Ken Olson. "Installation Basics for Solar Domestic Water Heating Systems," *Home Power* 94, 2003. The first in a series of three articles for those who would like to install their own solar hot water systems.

Marken, Chuck and Ken Olson. "One-Tank SDHW Storage with Electric Backup," *Home Power* 96, 2003.

All about retrofitting an existing electric water heater tank for solar hot water.

Marken, Chuck and Ken Olson. "DSHW Installation Basics. Part III: Drainback System," *Home Power* 97, 2003. Excellent reference for installers and do-it-yourselfers.

Mehalic, Brian. "Solar Hot Water Storage: Residential Tanks with Integrated Heat Exchangers," *Home Power* 131, pp. 70–78, 2009. Superb resource.

Mehalic, Brian. "Flat-plate and Evacuated-Tube: Solar Thermal Collectors," *Home Power* 132, 40–46, 2009.

Mehalic, Brian. "Drainback Solar Hot Water Systems," *Home Power* 138, 78–85, 2010. Valuable reading for those who want to learn about one of the most popular solar hot water systems on the market today.

Olson, Ken. "Solar Hot Water: A Primer," *Home Power* 84, 2001. Excellent overview of solar hot water systems and your options.

Olson, Ken. "Solar Hot Water, Homebrew Style," *Home Power* 88, 2002. For those who want to learn more about drainback systems.

Olson, Ken. "Solar Hot Water for Cold Climates, Part I: Closed Loop Antifreeze System Components," *Home Power* 85, 2001. For those interested in installing a solar hot water system in a climate where wintertime freezing is a regular occurrence.

Owens, Bob. "Florida Batch Water Heater," *Home Power* 93, 2003. For those interested in installing a solar batch heater.

Owens, Bob. "My Solar Heated Hot Tub," *Home Power* 105, 2005. For those interested in solar hot tubs, this article is a must.

Patterson, John and Suzanne Olsen. "Single-Tank Solar Water Systems," *Home Power* 124, 42–46, 2008. A detailed look at one of the important and economical options for a solar hot water system.

Perlin, John. "Solar Hot Water History," *Home Power* 100, 2004. A great overview of the history of solar hot water.

Sindelar, Allan and Phil Campbell-Graves. "How to Finance Your Renewable Energy Home," *Home Power* 103, 2004. Very useful article.

Sklar, Scott. "Selecting a Solar Heating System," *Solar Today* 18(5), 2004. A good look at the economics of solar hot water systems.

Sklar, Scott, and Kenneth Sheinkopf. *Consumer Guide to Solar Energy: More Ways to Reduce Your Energy Bills and Save the Environment*. Bonus Books, 1995. Delightful introduction to many different solar applications, including solar hot water.

Tonnessen, Roy W. "Solar Water Heating," *Home Power* 91, 2002. A case study worth your while to read.

Weaver, Jennifer. "Tankless Is In," *Home Power* 105, 2005. Excellent overview of tankless water heaters.

Wilson, Jib. "Hawaiian Heat," *Solar Today* 14(3), 2000. Production builders are adding domestic solar hot water systems with help of the state.

Magazines and Newsletters

Backwoods Home Magazine. P.O. Box 712, Gold Beach, OR 97444. Tel: (800) 835- 2418. Publishes articles on all aspects of self-reliant living, including renewable energy strategies such as solar. Website: backwoodshome.com.

The CADDET Renewable Energy Newsletter. 168 Harwell, Oxfordshire OX11 ORA, United Kingdom. Tel: +44 123335 432968. Quarterly magazine published by the CADDET Centre for Renewable Energy. Covers a wide range of renewable energy topics.

Earth Quarterly (formerly *Dry Country News*). Box 23-J, Radium Springs, NM 88054. Tel: (505) 526-1853. A new magazine devoted to living close to, and in harmony with nature. Covers all aspects of natural life including homebuilding and renewable energy. Website: zianet.com/earth.

EREN Network News. Newsletter of the Department of Energy's Energy-Efficiency and Renewable Energy Network. See listing under organizations.

Home Energy Magazine. 2124 Kittredge Street, No. 95, Berkeley, CA 94704. Great for those who want to learn more about ways to save energy in conventional home construction.

Home Power. P.O. Box 520, Ashland, OR 97520. Tel: (800) 707-6585. Publishes numerous extremely valuable how-to and general articles on renewable energy, including solar hot water, PVs, wind energy, microhydro electric, and occasionally an article or two on passive solar heating and cooling. This magazine is a gold mine of information, an absolute must for anyone interested in learning more. The magazine also contains important product reviews and ads for companies and professional installers. CDs containing back issues can be purchased from *Home Power* website: homepower.com.

Inside and Out. Newsletter of the Passive Solar Industries Council. See listing under organizations.

Mother Earth News. 1503 SW 42nd St., Topeka, KS 66609. One of my favorite magazines. Usually publishes a very useful article in each issue on some aspect of renewable energy. Website: motherearthnews.com.

Solar Today. ASES, 2400 Central Ave., Suite G-1, Boulder, CO 80301. Tel: (303) 443- 3130. This magazine, published by the American Solar Energy Society, contains lots of good information on passive solar, solar thermal, photovoltaics, hydrogen, and other topics, but not much how-to information. Also lists names of engineers, builders, and installers and lists workshops and conferences. Website: solartoday.org.

Organizations

American Solar Energy Society. 2400 Central Avenue, Suite G-1, Boulder, CO 80301. Publishes *Solar Today* magazine and sponsors an annual national meeting. Also publishes an online catalog of publications and sponsors the National Tour of Solar Homes. Contact this organization to find out about an ASES chapter in your area. Website: ases.org.

Center for Building Science. Lawrence Berkeley National Laboratory's Center for Building Science works to develop and commercialize energy-efficient technologies and to document ways of improving the energy efficiency of homes and other buildings while protecting air quality. Website: eetd.lbl.gov.

Center for Renewable Energy and Sustainable Technologies (CREST). 1612 K St. NW, Suite 410, Washington, D.C. 20006. Tel: (202) 293-2898. Nonprofit organization dedicated to renewable energy, energy efficiency, and sustainable living. Website: thinkenergy.org.

El Paso Solar Energy Association. P.O. Box 26384, El Paso, TX 79926. Active in solar energy, especially passive solar design and construction. Website: epsea.org/design.html.

Energy Efficiency and Renewable Energy Clearinghouse. P.O. Box 3048, Merrifield, VA 22116. Tel: (800) 363-3732. Great source of a variety of useful information on renewable energy.

Florida Solar Energy Center. FSEC, 679 Clearlake Road, Cocoa, FL 32922. Tel: (321) 638-1000. A research institute of the University of Central Florida Research and education on passive solar, cooling, and photovoltaics. Website: fsec.ucf.edu.

Midwest Renewable Energy Association. P.O. Box 249, Amherst, WI 54406. Tel: (715) 824-5166. Actively promotes solar energy and offers valuable workshops and a superb annual energy fair. Also sponsors a smashing annual event on the Summer Solstice, the Midwest Renewable Energy Fair. It's a must-do — many times! Website: the-mrea.org.

National Renewable Energy Lab. NREL, 1617 Cole Blvd., Golden, CO 80401. Tel: (303) 384-7349. Center for Buildings and Thermal Systems. Key players in research and education on energy efficiency and passive solar heating and cooling. Website: nrel.gov/buildings/highperformance.

North Carolina Solar Center. Box 7401, Raleigh, NC 27695. Tel: (919) 515-3480. Offers workshops, tours, publications, and much more. Website: www.ncsc.ncsu.edu.

Renewable Energy Training and Education Center. 1679 Clearlake Road, Cocoa, FL 32922. Tel: (407) 638-1007. Offers hands-on training and certification courses in the US and abroad for those interested in becoming certified in solar installation.

Solar Energy International. P.O. Box 715, Carbondale, CO 81623. Tel: (970) 963- 8855. Offers a wide range of workshops on solar energy, wind energy, and natural building. Website: solarenergy.org.

Solar Living Institute. P.O. Box 836, Hopland, CA 95449. Tel: (707) 744-2017. A nonprofit organization that offers frequent hands-on workshops on solar energy and many other topics. Be sure to tour their facility if you are in the neighborhood. Website: solarliving.org.

The Evergreen Institute: Center for Renewable Energy and Green Building, 3028 Pin Oak Road, Gerald, MO 63037. Tel: (303) 883-8290. The author's educational center, which offers hands-on workshops in solar hot water, home energy efficiency, passive solar, small wind energy, green building and many other topics. Be sure to tour the facility, but call first for a tour. Website: evergreeninstitute.org.

Chapter 4: Free Heat — Passive Solar and Heat Pumps

Passive Solar

Publications

Chiras, Daniel D. "Build a Solar Home and Let the Sunshine in," *Mother Earth News* 193, 2002. A survey of passive solar design principles and a case study showing the economics of passive solar heating.

Chiras, Dan. "Learning from Mistakes of the Past," *The Last Straw* 36, 2001. Describes common errors in passive solar design.

Chiras, Dan. "Passive Solar Retrofit," *Home Power* 138, 106–111, 2010. A detailed examination of ways to retrofit a home for passive solar.

Chiras, Daniel D., ed., "Solar Solutions," *The Last Straw* 36, 2001. Over a dozen articles, many by me, on passive solar heating, integrated design, thermal mass, and more.

Chiras, Dan. "Sunshine from a Tube," *Mother Earth News* 202, 2004. Brief introduction to solar tube skylights.

Chiras, Daniel D. *The Solar House: Passive Solar Heating and Cooling.* Chelsea Green, 2002. A detailed, readable guide for designing and building homes for passive solar heating and cooling.

Chiras, Dan. "Sun-Wise Design: Avoid Passive Solar Design Blunders," *Home Power* 105, 2005. Important look at the most costly and most common mistakes in passive solar design.

Crosbie, Michael. J., ed., *The Passive Solar Design and Construction Handbook.* John Wiley and Sons, 1997. A fairly technical manual on passive solar homes. Contains detailed drawings and case studies.

Crowther, Richard I. *Affordable Passive Solar Homes: Low-Cost Compact Designs.* SciTech Publishing, 1984. Contains some valuable background information on passive solar design and numerous designs for passive solar homes.

Energy Division, North Carolina Department of Commerce. *Solar Homes for North Carolina: A Guide to Building and Planning Solar Homes.* North Carolina Solar Center, 1999. Available online at the North Carolina Solar Center's website.

Galloway, Terry. *Solar House: A Guide for the Solar Designer.* Architectural Press, 2004. Fairly technical guide on a wide range of solar designs, including passive solar.

Gillett, Drew and Nick Pine. "Soldier's Grove Soldiers On," *Solar Today* 17(6), 2003. Inspiring look at a solar town in Wisconsin that uses passive and active solar technologies to provide heat to homes and stores.

Johnston, David and Kim Master. *Green Remodeling: Changing the World One Room at a Time.* New Society Publications, 2004. Many ideas that boost energy efficiency in existing homes and increase reliance on renewable energy, including passive solar.

Kriescher, Paul. "New England Style Passive Solar," *Solar Today* 14(3), 2000. An interesting case study in residential passive solar heating.

Kachadorian, James. *The Passive Solar House.* Chelsea Green, 1997. Presents a lot of good information on passive solar heating and an interesting design that has been fairly successful in cold climates.

Kubusch, Erwin. *Home Owner's Guide to Free Heat: Cut Your Heating Bills Over 50%.* Sunstore Farms, 1991. A self-published book with lots of good basic information.

Amanda Griscom. "Super Solar Homes Everyone Can Afford," *Mother Earth News* 207, 2005. Those interested in building a new home should check out this article.

Marken, Chuck. "Solar Hot Air System Design," *Home Power* 98, 2004. Valuable resource for those who can't retrofit for passive solar or solar hot water heating systems.

Marken, Chuck. "Solar Hot Air Systems, Part II," *Home Power* 99, 2004. Valuable resource.

Miller, Burke. *Solar Energy: Today's Technologies for a Sustainable Future.* American Solar Energy Society, 1997. Extremely valuable resource; contains numerous case studies showing how passive solar heating can be used in different climates, even in some solar-deprived places.

Moore, Steve and Carol Moore. "Winter Food Production in Pennsylvania — Without Fossil Fuels," *Home Power* 99, 2004. Learn to grow much of your own food year round without artificially heating a greenhouse.

Niklas, Mike. "High Performance Schools — It's a No-Brainer," *Solar Today* 16I (3), 2002. A fascinating look at the greening of a high school in Oregon, including energy efficiency, passive solar heating, passive cooling, and daylighting.

Olson, Ken and Joe Schwartz. "Home Sweet Solar Home: A Passive Solar Design Primer," *Home Power* 90, 2002. Superb introduction to passive solar design principles.

Passive Solar Industries Council. *Passive Solar Design Strategies: Guidelines for Home Builders.* PSIC, undated. Extremely useful book with worksheets for calculating a house's energy demand, the amount of backup heat required, the temperature swing one can expect given the amount of thermal mass you've installed, and the estimated cooling load. You can order a copy from the Sustainable Buildings Industry Council (SBIC, formerly the PSIC) with detailed information for your state, so you can design a home to meet the requirements of your site.

Reynolds, Michael. *Comfort in Any Climate.* Solar Survival Press, 1990. A brief but informative treatise on passive heating and cooling.

Riversong, Robert. "Designing a Passive Solar Slab," *Home Power* 136, 60–66, 2010. Very worthwhile reading for anyone serious about building a passive solar home.

Sklar, Scott, and Kenneth Sheinkopf. *Consumer Guide to Solar Energy: More Ways to Reduce Your Energy Bills and Save the Environment.* Bonus Books, 1995. Delightful introduction to many different solar applications, including passive solar heating.

Solar Survival Architecture. "Thermal Mass vs. Insulation." *Earthship Chronicles.* Solar Survival Architecture, 1998. Basic treatise on passive solar heating and cooling.

Sustainable Buildings Industry Council. *Designing Low-Energy Buildings: Passive Solar Strategies and Energy-10 Software.* SBIC, 1996. A superb resource! This book of design guidelines and the *Energy-10* software that comes with it enables builders to analyze the energy and cost savings in building designs. Helps permit region-specific design.

Taylor, John S. *Shelter Sketchbook: Timeless Building Solutions.* Chelsea Green, 1983. Pictorial history of building that will open your eyes to intriguing design solutions to achieve comfort, efficiency, convenience, and beauty.

Videos

Buildings for a Sustainable America. A concise overview of passive solar buildings and their benefits. Available from the Sustainable Buildings Industry Council (SBIC), 1331 H Street NW, Suite 1000, Washington, D.C. 20005. Tel: (202) 628-7400. Website: sbicouncil.org.

Organizations

Sustainable Buildings Industries Council. SBIC, 331 H. Street NW, Suite 1000, Washington, D.C. 20005. Tel: (202) 628- 7400. This organization has a terrific website with information on workshops, books, and publications on energy-efficient, passive solar design, and links to many other international, national, and state solar energy organizations. Publishes a newsletter, *Buildings Inside and Out.* Website: psic.org.

Midwest Renewable Energy Association: see listing in Chapter 3

Solar Energy International: see listing in Chapter 3.

Solar Living Institute: see listing in Chapter 3.

The Evergreen Institute: see listing in Chapter 3.

Heat Pumps

Publications

Malin, Nadav, and Alex Wilson. "Ground-Source Heat Pumps: Are They Green?" *Environmental Building News* 9(1), 2000. Detailed overview on ground-source heat pumps.

National Renewable Energy Lab. "Geothermal Heat Pumps," published online at eere.energy.gov/geothermal/heat pumps. Great overview of ground-source heat pumps.

Persons, Jeff. "The Big Dig," *Mother Earth News* 185, 102. A brief introduction to ground-source heat pumps.

Organizations

Geo-Heat Center, Oregon Institute of Technology, 3201 Campus Dr., Klamath, OR. 97601. Tel: (541) 885-1750. Technical information on heat pumps. Website: geoheat.oit.edu.

Geothermal Heat Pump Consortium, Inc. 701 Pennsylvania Ave, NW, Washington, D.C. 20004-2696. Tel: (888) 333-4472. General and technical information on heat pumps. Website: ghpc.org.

International Ground Source Heat Pump Association. 490 Cordell South, Stillwater, OK. 74078-8018.

Tel: (405) 744-5175. Provides a list of equipment manufacturers, installers by state, and numerous other resources for contractors, homeowners, students, and the general public. Website: igshpa.okstate.edu.

US Department of Energy, Office of Geothermal Technologies. EE-12, 1000 Independence Avenue, S.W., Washington, D.C. 20585-0121. Tel: (202) 586-5340. Carries out research on GSHPs and works closely with industry to implement new ideas.

Chapter 5: Solar Hot Air and Hot Water Systems: Affordable Heat from the Sun

Publications

Gillett, Drew and Nick Pine. "Soldier's Grove Soldiers On," *Solar Today* 17(6), 2003. An inspiring look at a solar town in Wisconsin that uses passive and active solar technologies to provide heat to homes and stores.

Hyatt, Rod. "Hydronic Heating on Renewable Energy," *Home Power* 79, 2000. Provides a lot of practical advice on building your own radiant floor heating system and powering it with photovoltaic panels.

Kriescher, Paul. "Energy Performance on a Budget," *Solar Today* 14(5), 2000. A fascinating case study in residential active and passive solar heating in New Hampshire. A must read.

Lane, Tom, 2004, *Solar Hot Water Systems: Lessons Learned 1977 to Today*. Energy Conservation Services. The definitive solar technical manual.

Marsden, Guy. "Solar Heat for My Main Workshop," *Home Power* 89, 2002. A case study worth reading by those interested in using solar hot water to provide space heat.

Pine, Nick. "Solar Heat in Snow Country," *Solar Today* 17 (1), 2003. An inspiring story about active solar heating in a US Customs border station in Vermont.

Ramlow, Bob and Benjamin Nusz. *Solar Water Heating — Revised & Expanded Edition: A Comprehensive Guide to Solar Water and Space Heating*. New Society Publishers, 2010. Designed for a readership ranging from the curious homeowner to the serious student or professional.

Simpson, Walter. "Adventure in Solar Living," *Solar Today* 17(5), 2003. An inspiring tale of a passive solar/solar hot water heated home in Buffalo, New York.

Chapter 6: Wood Heat

Publications

Barden, Albert A. *AlbiCore Construction Manual*. Maine Wood Heat Company, 1996. Detailed construction manual for one type of masonry heater.

Barden, Albert A. *The Finnish Fireplace: Construction Manual*. Maine Wood

Heat Company, Inc., 1984. The only complete English language primer on making masonry heaters. Available through the Maine Wood Heat Company.
Website: mainewoodheat.com.

Barden, Albert A. and Keikki Hyytiainen. *Finnish Fireplaces: Heart of the Home*. Building Book Ltd., 1988. A valuable resource for anyone wanting to learn more about Finnish masonry stoves. Available through the Maine Wood Heat Company listed above.

British Columbia Ministry of Environment, Land, and Parks. "Reducing Wood Stove Smoke: A Burning Issue," Sept. 1994. Website: env.gov.bc.ca/epd/epdqa/ar/particulates/rwssabi.html

Gulland, John. "The Art of the Wood Cookstove," *Mother Earth News* 207, 2005. A detailed look at wood cookstoves.

Gulland, John. "Woodstove Buyer's Guide," *Mother Earth News* 189, 2002. Superb overview of wood stoves with a useful table to help you select a model that meets your needs.

Gulland, John. "Responsible Wood Heating: A Kind-to-the-Environment Guide," *Home Power* 99, 2004. For those who want to burn wood responsibly, that is, with minimal impact, this article is a must.

Johnson, Dave. *The Good Woodcutter's Guide: Chain Saws, Portable Sawmills, and Woodlots*. Chelsea Green, 1998. A practical guide to felling trees and cutting firewood safely.

Lyle, David. *The Book of Masonry Stoves: Rediscovering an Old Way of Warming*. Chelsea Green, 1984. This book contains a wealth of information on the history, function, design, and construction of masonry stoves.

Mink, Kate. "Living with a Masonry Stove," *Home Power* 103, 2004. A personal account of what it is like to live with a masonry heater; well worth reading.

Pahl, Greg. "Wood-Fired Central Heat," *Mother Earth News* 196, 2003. For additional information on wood furnaces.

Organizations

Hearth, Patio, and Barbecue Association. (Formerly the Hearth Products Association.) 1601 North Kent Street, Suite 1001, Arlington, VA. 22209. International trade association that promotes the interests of the hearth products industry. Offers lots of valuable information. Website: hpba.org.

Masonry Heater Association of North America. 1252 Stock Farm Road, Randolph, VT. 05060. Tel: (802) 728-5896. Publishes a valuable newsletter and has a website with links to dealers and masons who design and build masonry stoves. Website: mha-net.org.

Wood Heat Organization. 410 Bank Street, Suite 117, Ottawa, Ontario Canada K2P 1Y8. Promotes safe, responsible use of wood for heating. Website: woodheat.org.

Chapter 7: Passive Cooling: Staying Cool Naturally

Publications

Chiras, Dan. "Sunshine from a Tube," *Mother Earth News* 202, 2004. A brief introduction to solar tube skylights, a device that helps reduce heat gain in the summer.

Mossberg, Cliff. "Passive Cooling. Part 1– Basic Principles," *Home Power* 82, 2001. A fairly technical, but very valuable resource on passive cooling.

Mossberg, Cliff. "Passive Cooling. Part 2 – Applied Construction," *Home Power* 82, 2001. Part 2 of the piece described above.

Galloway, Terry. *Solar House: A Guide for the Solar Designer*. Architectural Press, 2004. Fairly technical guide on a wide range of solar designs, including passive cooling.

Givoni, Baruch. *Passive and Low Energy Cooling of Buildings*. John Wiley and Sons, 1994. A fairly technical book, but one of the few resources on the subject.

Niklas, Mike. "High Performance Schools — It's a No-Brainer," *Solar Today* 16 (3), 2002. A fascinating look at the greening of a high school in Oregon, including energy efficiency, passive cooling, and daylighting.

Pagés, Fernando. *Building an Affordable House*, Taunton Press, 2005.

NOTE: *See resources for passive solar heating, as many concepts and techniques that make passive solar heating work, also function to passively cool a home.*

Chapter 8: Solar Electricity — Powering Your Home with Solar Energy

Publications

Berton, John. "Off-Grid in Chicago," *Home Power* 80, 2001. For those who are skeptical about solar electricity in a city, this is a must read.

Butti, Ken and John Perlin. *Golden Thread: 2500 Years of Solar Architecture and Technology*. Cheshire Books, 1980. Delightful history of solar energy.

Dankoff, Windy. "Solar Water Pumping Makes Sense," *Home Power* 97, 2003. Great primer by the nation's leading expert on the subject.

Davidson, Joel. *The New Solar Electric Home: The Photovoltaics How-To Handbook*. Aatec Publications. Comprehensive and highly readable guide to photovoltaics, a brand new edition.

Del Vecchio, David. "Optimizing a PV Array," *Home Power* 130, 52–56, 2009. A must read for homeowners and future installers.

Ewing, Rex A. *Power With Nature: Solar and Wind Energy Demystified.* PixyJack Press, 2003. Skip to the second half if you want to get to the meat of the matter.

Galocy, Baran. "Build Your Own Adjustable PV Mount," *Home Power* 97, 2003. Instructions on how to build an adjustable PV rack.

Goodnight, Jim. "Sizing a Generator for Your RE System," *Home Power* 138, 88–91, 2010. A fairly technical piece for those who want to learn more about backup generators for off-grid systems.

Gourley, Colleen. "Production Builders Go Solar," *Solar Today* 16(1), 2002. Inspiring story about the incorporation of solar electricity into homes in California by large-scale production builders.

Guevara-Stone, Laurie and Ian Woofenden. "Choosing Your RE Installer," *Home Power* 127, 48–52, 2008. Read this article before you start calling potential installers.

Hackleman, Michael and Claire Anderson. "Harvest the Wind," *Mother Earth News,* June/July, 2002. A wonderful introduction to wind power.

Hren, Rebekah. "A Peek Inside PV," *Home Power* 132, 58–64, 2009. For those interested in learning more about the structure and function of PV cells.

Kazmerski, Lawrence L. "Photovoltaic Myths — The Seven Deadly Sins," *Solar Today* 16 (4), 2002.

Kelin, Ken and Paul Hanley. "Winter Cattle Watering with Automated Solar Pumps," *Home Power* 98, 2004. Learn more about ways to pump water with solar electricity.

Kohn, Lin. "Solar in San Francisco," *Solar Today* 1(4), 2003. A great look at what one progressive city is doing to increase its reliance on renewable energy.

Komp, Richard J. *Practical Photovoltaics: Electricity from Solar Cells.* 3rd ed. Aatec Publications, 1999. Fairly popular book on PVs.

LaForge, Christopher. "Choosing the Best Batteries: 2009 Battery Specifications Guide," *Home Power* 127, 80–88, 2009. Superb piece for those interested in a battery-based system.

Larson, Kelly. "Off-Grid Inverter Buyer's Guide," *Home Power* 132, 96–103, 2009. Compares all the major models of off-grid inverters. Great shopping guide.

Lensch, Erik. "Pumping Water with Sunshine," *Home Power* 125, 74–78, 2008. Worthwhile reading for those who live on farms with livestock.

Linkous, Clovis A. "Solar Energy Hydrogen — Partners in a Clean Energy Economy." *Solar Today* 13(4), 1999. A detailed and somewhat technical article on hydrogen production using solar electricity.

Marsden, Guy. "Microinverters Make a Simple DIY Installation," *Home Power* 136, 50–56, 2010. This article explains some of the basics of microinverters and how they simplify a PV installation.

Mayfield, Ryan. "Rack and Stack: PV Array Mounting Options," *Home Power* 124, 58–64, 2008. Looks at all the options for mountain a PV array on a rack.

Mayfield, Ryan. "Grid-Tied Inverter Buyer's Guide," *Home Power* 138, 58–66, 2010. Great comparison of grid-tied inverter features.

NREL. *The Colorado Consumer's Guide to Buying a Solar Electric System*. National Renewable Energy Lab, 1998. Provides basic information about purchasing, financing, and installing photovoltaic systems in Colorado that is applicable to many other states and countries as well. Contact NREL's Document Distribution Service at: (303) 275-4363 for a free copy.

NREL. *The Borrower's Guide to Financing Solar Energy Systems*. National Renewable Energy Lab, 1998. Provides information about nationwide financing programs for photovoltaics and passive solar heating. Contact NREL's Document Distribution Service at (303) 275-4363 for a free copy.

Pahl, Greg. "Choosing a Backup Generator," *Mother Earth News* 202, 2004. Fairly detailed overview of what to look for when buying a backup electrical generator for an RE system.

Peavy, Michael A. *Fuel from Water: Energy Independence with Hydrogen*, 8th ed. Merit, Inc., 1998. Technical analysis for engineers and chemists.

Perez, Richard. "To Track ... or Not to Track," *Home Power* 101, 2004. Great little piece on tracking devices for PVs and whether they make sense for your site.

Perez, Richard. "Flooded Lead-Acid Battery Maintenance," *Home Power* 98, 2004. Read and memorize this, and put the advice into practice if you plan on installing a stand-alone system.

Perlin, J. *From Space to Earth: The Story of Solar Electricity*. Aatec Publications, 1999. A wonderfully readable history of solar electricity.

Perlin, John, Lawrence Kazmerski, and Susan Moon. "Good as Gold: The Silicon Solar Cell Turns 50," *Solar Today* 28(1), 2004. A brief history of PV research and development.

Pinkham, Linda. "From the Ground Up: My RE System Design Choice," *Home Power* 106, 2005. A personal account that walks readers through the issues one woman faced when considering a solar electric system.

Radoff, Joshua H. "Solar in the City," *Solar Today* 18(5), 2004. An interesting look at solar electric installations in New York City.

Rousso, Mo. "Get Solarized Now: Financing your PV System," *Home Power* 129, 58–62, 2009. Important reading for those who can't afford to pay for a PV system outright.

Rastelli, Linda. "Energy Independence With All the Comforts," *Solar Today* 16(1), 2002. An inspiring story about a passive solar/solar electric home in Washington, D.C.

Roberts, Simon. *Solar Electricity: A Practical Guide to Designing and Installing Small Photovoltaic Systems.* Prentice-Hall, 1991. Good reference, but a bit dated.

Russell, Scott. "Solar-Electric Systems Simplified," *Home Power* 104, 2005. Good, solid, and well-illustrated introduction to the components of a solar electric system.

Russell, Scott. "Starting Smart: Calculating Your Energy Appetite," *Home Power* 102, 2004. Great little introduction to household load analysis to determine household electrical demand.

Sanchez, Justine. "PV Energy Payback," *Home Power* 127, 32–36, 2008. Looks at the energy it takes to make PV systems and how rapidly a PV system generates that same amount of energy.

Sanchez, Justine. "Tracked: PV Array Systems and Performance," *Home Power* 131, 50–56, 2009. Great piece on the economics of trackers.

Sanchez, Justine. "Solar Site Assessment," *Home Power* 130, 46–50, 2009. For those interested in getting into the solar business.

Sanchez, Justine. "Sizing Batteryless Grid-Tied PV Arrays," *Home Power* 138, 62–67, 2010. Read this piece to learn how to size a grid-connected system.

Sanchez, Justine and Brad Burritt. "Charge Controller Buyer's Guide," *Home Power* 129, 72–77, 2009. Another terrific shopping tool for those interested in battery-based PV systems.

Schaeffer, John and the Real Goods staff. *Solar Living Source Book*, 10th ed., Real Goods, 1999. Contains an enormous amount of background information on wind, solar, and microhydroelectric.

Schwartz, Joe. "What's Going On — The Grid?" *Home Power* 106, 2005. An excellent look at inverters for grid-connected systems.

Schwartz, Joe. "Go Configure: Configuring Your PV Array," *Home Power* 87, 2002. Guide to wiring PV arrays. Very useful.

Seuss, Terri and Cheryl Long. "Eliminate Your Electric Bill: Go Solar, Be Secure," *Mother Earth News* 190, 2002. Excellent discussion of solar roofing materials.

Sindelar, Allan and Phil Campbell-Graves. "How to Finance Your Renewable Energy Home," *Home Power* 103, 2004. Very useful article.

Sindelar, Allan. "Engine Generator Basics," *Home Power* 131, 96–102, 2009. A must read for anyone interested in going off grid and thinking about installing a gen-set with their PV or wind system.

Solar Energy International. *Photovoltaic Design Manual, Version 2.* Solar Energy International. A manual on designing, installing, and maintaining a PV system. Used in SEI's PV design and installation workshops.

Stockman, Douglas. "The Hard Part About Wind Turbines," *Home Power* 86, 2002. Personal tale about the difficulties of siting a wind machine in a rural area.

Strong, Steven and William G. Scheller. *The Solar Electric House: Energy for the Environmentally Responsive, Energy-Independent Home.* Sustainability Press, 1993. Comprehensive and technical guide to solar electricity.

Symanski, Paul. "Money from the Sun: An Investor's Guide to Solar-Electric Profits," *Home Power* 100, 2004. Very interesting comparison of the economic payback of grid-connected solar electric systems in Arizona to an investment in the stock market.

Taylor, Jeremy. "Pump up the Power," *Home Power* 127, 72–77, 2008. How to maximize electrical production from a PV system.

Wiles, John. "Microinverters and AC PV Modules," *Home Power* 136, 112–113, 2010. Great little piece on a new and exciting inverter.

Woofenden, Ian. "Battery Filling Systems of the Americas: Single-Point Watering System," *Home Power* 100, 2004. This article is a must for those who would like to reduce battery maintenance.

Woofenden, Ian. "Off or On Grid? Getting Real," *Home Power* 128, 40–45, 2008. A piece everyone who's thinking about installing an off-grid system should read.

Yerkes, Bill. "40 Years of Solar Power," *Solar Today* 18(1), 2004. A historical survey of PVs from an insider.

Yewdall, Zeke. "PV Orientation," *Home Power* 93, 2003. A must read for those with less-than-perfect solar sites.

Videos

An Introduction to Residential Solar Electricity with Johnny Weiss. Good basic introduction to solar electricity. Produced by Scott S. Andrews, P.O. Box 3027, Sausalito, CA 94965. Tel: (415) 332-5191. Also available through Gaiam Real Goods.

An Introduction to Solar Water Pumping with Windy Dankoff. A very useful introduction to the subject. Produced by Scott S. Andrews, P.O. Box 3027, Sausalito, CA 94965.

Tel: (415) 332-5191. Also available through Gaiam Real Goods.

An Introduction to Storage Batteries for Renewable Energy Systems with Richard Perez. This is one of the best videos in the series. It's full of great information. Produced by Scott S. Andrews, P.O. Box 3027, Sausalito, CA 94965. Tel: (415) 332-5191. Also available through Gaiam Real Goods.

The Solar-Powered Home with Rob Roy. An 84-minute video that examines basic principles, components, set-up, and system planning for an off-grid home featuring tips from America's leading experts in the field of home power. Can be purchased from the Earthwood Building School at 366 Murtagh Hill Road, West Chazy, NY, 12992. Tel: (518) 493-7744. Website: cordwoodmasonry.com.

Organizations

American Solar Energy Society: see listing in Chapter 3.

Centre for Alternative Technology. Address: Machynlleth, Powys SY20 9Az., UK. Tel: 01654 703409. This educational group in the United Kingdom offers workshops on alternative energy, including wind, solar, and microhydroelectric. Website: cat.org.uk.

Center for Renewable Energy and Sustainable Technologies: see listing in Chapter 3.

Midwest Renewable Energy Association: see listing in Chapter 3

Solar Energy International: see listing in Chapter 3.

Solar Living Institute: see listing in Chapter 3.

The Evergreen Institute: see listing in Chapter 3.

Chapter 9: Wind Power — Meeting Your Needs for Electricity
Publications

Bartmann, Dan and Dan Fink. "Homebrew Wind Power: A Hands-On Guide to Harnessing the Wind," Buckville Publications: Masonville, CO, 2009. For those interested in building a high-quality wind turbine.

Butler, Roy and Ian Woofenden. "Wind-Electric System Maintenance," *Home Power* 135, 98–103, 2010. Be sure to read this before you purchase a wind system — so you know what you are getting into.

Ewing, Rex A. *Power With Nature: Solar and Wind Energy Demystified.* PixyJack Press, 2003. Skip to the second half if you want to get to the meat of the matter.

Green, Jim. "Small Wind Installations in Colorado," *Solar Today* 14 (1), 2000. Several case studies of wind installations.

Gipe, Paul. "Small Wind Systems Boom," *Solar Today* 1 (2), 2002. Good overview of small wind energy systems by one of America's leading experts on the subject.

Gipe, Paul. *Wind Power for Home and Business: Renewable Energy for the 1990s and Beyond.* Chelsea Green, 1993. Somewhat technical introduction to small and medium-sized wind generators.

Gipe, Paul. *Wind Energy Basics: A Guide to Small and Micro Wind Systems.* Chelsea Green, 1999. A brief introduction to wind energy for newcomers.

Hackleman, M. and C. Anderson. "Harvest the Wind." *Mother Earth News* 192, 2002. An easy-to-read and fact-filled article on wind power.

Kuebeck, Sr., Peter and Peter Kuebeck, Jr. "Old Jacobs — Current Again," *Home Power* 89, 2002. Personal story about one of the best wind machines on the market.

LaForge, Christopher. "Choosing the Best Batteries: 2009 Battery Specifications Guide," *Home Power* 127, 80–88, 2009. Superb piece for those interested in a battery-based system.

Osborn, D. "Winds of Change." *Solar Today* 17(6), 2003. Looks at a number of important issues, including bird kills by wind towers compared to other sources.

Pahl, Greg. "Choosing a Backup Generator," *Mother Earth News* 202, 2004. A fairly detailed overview of what to look for when buying a backup electrical generator for an RE system.

Pearen, Craig. "Brushless Alternators," *Home Power* 97, 2003. Brief but interesting look at brushless (low-maintenance) alternators for wind machines.

Piggott, Hugh. "Estimating Wind Energy," *Home Power* 102, 2004. A brief piece that will help you determine how much electrical energy you can produce on your site.

Perez, Richard. "Flooded Lead-Acid Battery Maintenance," *Home Power* 98, 2004. Read and memorize this, and put the advice into practice if you plan on installing a stand-alone system.

Preus, Robert. "Thoughts on VAWTs: Vertical Axis Wind Generator Perspectives," *Home Power* 104, 2005.

Raichle, Brian and Brent Summerville. "How Tall is Too Tall," *Home Power* 126, 84–89, 2008. Important reading on an extremely important topic: tower height for wind turbines.

Rastelli, Linda. "Energy Independence — With All the Comforts," *Solar Today* 16(1), 2002. The story of a mass-marketed passive solar/solar electric home.

Russell, Scott. "Starting Smart: Calculating Your Energy Appetite," *Home Power* 102, 2004. Great introduction to household load analysis (to determine household electrical demand).

Sagrillo, Mick. "Being Your Own Utility," *Solar Today* 17(5), 2003. For those who are considering going off grid.

Sagrillo, Mick. "On Intimate Terms with a Wind Generator," *Solar Today* 17(1), 2003. An inspiring tale about one do-it-yourselfer and his wind machine.

Sagrillo, Mick and Ian Woofenden. "How to Buy a Wind Generator," *Home Power* 131, 2009. The most valuable resource you will find on selecting a wind generator. Contains a chart showing details on many small wind turbines.

Short, Walter and Nate Blair. "The Long-Term Potential of Wind Power in the US" *Solar Today* 17(6), 2003. Important study of wind power's potential.

Sindelar, Allan and Phil Campbell-Graves. "How to Finance Your Renewable Energy Home," *Home Power* 103, 2004. Very useful article.

Swezey, Blair and Lori Bird. "Businesses Lead the 'Green Power' Challenge," *Solar Today* 15(1), 2001. An interesting look at what some major corporations like Toyota are doing to increase their reliance on renewable energy.

Swezey, Blair and Lori Bird. "Buying Green Power — You Really Can Make a Difference," *Solar Today* 17(1), 2003. An in-depth look at ways homeowners can tap into renewable energy (such as green tags) without installing a system on their home.

Woofenden, Ian. "Wind Generator Tower Basics," *Home Power* 105, 2005. Excellent overview of an important subject.

Woofenden, Ian. "A Beginner's Guide to Tower Climbing Safety," *Home Power* 128, 66–70, 2009. Read this if you are thinking about installing a climbable tower.

Woofenden, Ian. "A Second Wind," *Home Power* 130, 84–88, 2009. A story that should help anyone interested in a wind system.

Woofenden, Ian. "Battery Filling Systems of the Americas: Single-Point Watering System," *Home Power* 100, 2004. This article is a must for those who would like to reduce battery maintenance.

Woofenden, Ian. "Wind Power Curves: What's Wrong, What's Better," *Home Power* 127, 92–95, 2008. Very valuable article for those seriously thinking about installing a wind turbine.

Woofenden, Ian. *Wind Power for Dummies.* Wiley: New York, 2009. Fairly comprehensive guide to wind energy.

Videos

An Introduction to Residential Wind Power with Mick Sagrillo. Produced by Scott S. Andrews, P.O. Box 3027, Sausalito, CA 94965. Tel: (415) 332-5191.
A very informative video, especially for those wishing to install a medium-sized system.

Organizations

American Wind Energy Association. 122C Street, NW, Suite 380, Washington, D.C. 20001. Tel: (202) 383-2500. This organization also sponsors an annual conference on wind energy. The website contains a list of publications, an online newsletter, frequently asked questions, news releases, and links to companies and organizations. Website: awea.org.

British Wind Energy Association. 26 Spring Street, London W2 1JA. Tel: 0171 402 7102. Actively promotes wind energy in Great Britain. Check out the website for fact sheets, answers to frequently asked questions, links, and a directory of companies. Website: bwea.com.

Centre for Alternative Technology. Machynlleth, Powys SY20 9Az., UK. Tel: 01654 703409. This educational group in the United Kingdom offers workshops on alternative energy, including wind, solar, and microhydroelectric. Website: cat.org.uk.

Center for Renewable Energy and Sustainable Technologies. See listing in Chapter 3.

European Wind Energy Association. Rue du Trone 26, B-1000, Brussels, Belgium. Tel: +32 2 546 1940. Promotes wind energy in Europe. The organization publishes the *European Wind Energy Association Magazine*.

The website contains information on wind energy in Europe and offers a list of publications and links to other sites. Website: ewea.org.

National Wind Technology Center of the National Renewable Energy Laboratory. 1617 Cole Blvd., Golden, CO 80401-3393. Tel: (303) 275-3000. The website provides a search mode so you can check out the site, and provides a great deal of information on wind energy, including a wind resource database. Website: nrel.gov/wind.

Real Goods Solar Living Institute: see listing in Chapter 3.

Midwest Renewable Energy Association: see listing in Chapter 3

Solar Energy International: see listing in Chapter 3.

Solar Living Institute: see listing in Chapter 3.

The Evergreen Institute: see listing in Chapter 3.

Chapter 10: Microhydropower: Generating Electricity from Running Water

Publications

Aronson, Jeffe. "Choosing Microhydro ... Clean Electricity in the Outback," *Home Power* 101, 2004. Valuable reading for anyone interested in low-head microhydro.

Gardner, Ken and Ian Woofenden. "Hydro-electric Turbine Buyer's

Guide," *Home Power* 136, 100–108, 2010. Excellent resource that will save you a lot of time if you are shopping for a microhydro turbine.

Greacen, Chris. "Low-Head Microhydro Thai Style," *Home Power* 124, 76–80, 2008. Looks at one of the often over-looked options for microhydro.

Isley, Bill. "Happy on One Kilowatt," *Mother Earth News* 180, 92, 2000. A story of one couple's quest for energy independence through microhydro.

Maxwell, Steve. "Homestead Hydropower," *Mother Earth News* 208, 2005. A good overview of the subject.

New, Dan. "Intro to Hydropower. Part 1: Systems Overview," *Home Power* 103, 2004. The first piece in a superb series.

New, Dan. "Intro to Hydropower. Part 2: Measuring Head and Flow," *Home Power* 104, 2004. The second piece in the series; this one covers important information on assessing a site.

New, Dan. "Intro to Hydropower. Part 3: Power, Efficiency, Transmission & Equipment Selection," *Home Power* 105, 2005. The final piece in the series.

Ostermeier, Jerry. "Microhydro Intake Design," *Home Power* 124, 68–73. Detailed look at the various types of intake structures for microhydro systems.

Ostermeier, Jerry. "Pipeline: Hydro-Electric Penstock Design," *Home Power* 125, 56–73, 2008. Detailed look at the various types of intake structures for microhydro systems.

Ostermeier, Jerry and Joe Schwartz. "The Electric Side of Hydro Power: Power Transmission and Regulation Considerations," *Home Power* 126, 68–73, 2008. A must read for anyone interested in installing microhydro.

Pahl, Greg. "Choosing a Backup Generator," *Mother Earth News* 202, 2004. A fairly detailed overview of what to look for when buying a backup electrical generator for an RE system.

Perez, Richard. "Flooded Lead-Acid Battery Maintenance," *Home Power* 98, 2004. Read and memorize this, and put the advice into practice if you plan on installing a stand-alone system.

Russell, Scott. "Starting Smart: Calculating Your Energy Appetite," *Home Power* 102, 2004. Great introduction to household load analysis (to determine household electrical demand).

Woofenden, Ian. "Battery Filling Systems of the Americas: Single-Point Watering System," *Home Power* 100, 2004. This article is a must for those who would like to reduce battery maintenance.

Woofenden, Ian. "Hydro Design Considerations," *Home Power* 132, 78–132, 2009. Valuable reading for anyone interested in learning more about microhydro.

Videos

An Introduction to Residential Microhydro Power with Don Harris. Produced by Scott S. Andrews. P.O. Box 3027, Sausalito, CA 94965. Tel: (415) 332-5191. Outstanding video packed with lots of useful information.

Index

A

AC electricity: and microhydro systems, 270–271, 281, 285–286, 287; and solar electricity, 203–204; 240 volt, 232

AGM mat batteries, 236

air conditioners, 5, 192–193, 194

air sealing, 38, 39

air-source heat pumps, 129, 130

amorphous silicon technology, 228, 229

anti-islanding, 219

antifreeze (propylene glycol), 93, 95–100, 145

appliances: assessing power consumption of, 220–221, 222; and modified sine wave electricity, 230; and passive cooling, 175, 176, 177–178; and retrofits, 61–62, 68–72; running on DC power, 214; 240-volt, 232

arbors for shade, 181–182

attics, 53–54, 179, 189–191

awnings, 183

B

backdraft, 160

backdraft dampers, 135, 141

baffles, 156

batch hot water systems, 85–88, 89–90

batteries: buying, 232–235; maintenance of, 214–218, 234, 236; matched to inverters, 229–230, 231; in microhydro systems, 281, 286, 287–288; role in solar electric systems, 210, 212; and safety, 215, 234–235; sealed, 235–237; sizing, 235

Becquerel, Edmond, 226

biomass, 14

blackouts, 218–219, 246

blankets, water heater, 65

blower doors, 43–45

boilers, 62, 64
bridging loss, 52
brownouts, 218–219
building codes, 148, 158
businesses, 24, 26, 139, 142, 199, 267

C
Cansolair's Solar Max, 138
car batteries, 233–234
carbon capture, 11–12
carbon dioxide emissions, 11–12, 41, 128, 159–160, 168
Carter, Jimmy, 82
catalytic converters/burners, 154–155
caulks, 48–49
ceilings, 53–56, 121, 161
CFLs (compact fluorescent lightbulbs), 33–36, 66–67
charge controllers, 210, 212, 230
clothes dryers, 177
CO_2 emissions, 11–12, 41, 128, 159–160, 168
coal, 10, 11–12, 16
combustion wood stoves, 153
community gardens, 29
compact fluorescent lightbulbs (CFLs), 33–36, 66–67
computers, 177
concrete, storage of heat in, 146
contractors, 237–238
controllers, 288–289
cookers, solar, 122
cooling systems, 5, 55, 64, 112, 192–193, 194. *see also* passive cooling

cracks, sealing, 48–49, 185, 192.
  see also leaks, plugging
creosote, 152, 171
Culpepper, Kara, 1–2

D
daylighting, 112
DC electricity: and microhydro systems, 270–271, 280–281, 285, 286, 287; and solar electricity, 203–204, 214
disconnects, 205, 214, 218, 219
dishwashers, 68–69, 178
domestic solar hot water systems.
  see solar hot water systems
double-wall technique, 51–53
drainback systems, 97–99, 144
driveways, solar, 81, 82
dump loads, 146, 289

E
economics: of batteries, 218; of connection to grid, 198; of cooling systems, 55; of electricity, 64; of energy efficiency, 2–3; of masonry heaters, 172; of microhydro systems, 269, 292; of passive solar, 109, 125–126; of solar electric systems, 32, 33, 197–200, 220, 238; of solar hot air space heating systems, 140–141; of solar hot water space heating systems, 147; of solar hot water systems, 105–107; of water heating systems, 75; of wind power, 262–263, 264–265; of wood stoves, 160
electrical grid, 198, 204, 205–208, 210.
  see also grid–connected systems

electrical inspectors, 213–214

electrical outlets, leaks by, 45, 49

electricity (*see also* solar electric systems): assessing electrical load, 220–223, 272–273; cost of, 64; and ghost loads, 68–72; history of development, 242; managing power consumption, 274–275; measuring, 19–21; reducing use of, 175; 240-volt, 232

energy: conversion of, 15–17; definition of, 18; forms of, 14–15; measuring, 18–21

energy audits, 42–47, 48

energy conservation: benefits of, 39–40; future of, 72–73; and home insulation, 38–39; ideas for, 29–30, 33; incentives for, 39–40, 72; and quality of life, 36–37

energy consumption: assessing, 220–221, 222; of homes, 40–41; impact on environment, 41; managing, 274–275

energy efficiency: and appliances, 68–72; benefits of, 39–40; and cutting cost of renewable system, 37; economics of, 2–3; of ground-source heat pumps, 127–128; ideas for, 12–13, 28–30; and lightbulbs, 33–36, 66–68; and new homes, 42; and quality of life, 36–37; and water heaters, 101

environmental impact, 13, 25, 50, 280. *see also* carbon dioxide emissions; pollution

evaporative coolers, 192, 193

F

fans, 189–191, 193–194, 214

financial savings. *see* economics

fire hazards, 162, 163, 164

fireplaces, 149–150, 159

flashing, 55

flat plate collectors, 91–92

flow, measuring, 278–280

food, 9–10, 29

fossil fuels, 10, 11, 13, 14–15, 16, 26, 242–243

Fritts, E. E., 226

furnaces, 62, 64, 144, 162–164, 178

G

gardens, home, 29

gel cell batteries, 236, 237

generators: and hybrid systems, 247; in microhydro systems, 285–286; and solar electricity, 217, 220; in wind turbines, 245, 249–250

geothermal systems, 63, 126–129

GFIs (ground fault interrupters), 71

ghost loads, 69–72

golf cart batteries, 234

grants, 199, 263

green power, 263–264, 265

green tags, 264

grid-connected microhydro systems, 274

grid-connected PV systems: and buying an inverter, 229–230; maintenance of, 215; pros and cons of, 218–220; set up of, 200–208; sizing, 221

grid-connected systems with battery backup, 208–210, 215, 219, 246

grid-connected wind power systems, 246, 262

ground fault interrupters (GFIs), 71
ground-source heat pumps, 126–129

H
head, measuring, 275–278
heat exchangers, 79, 96
heat pumps: air-source, 129, 130, 193;
   ground-source, 126–129; how they
   work, 62–64
home owner association rules, 83–84, 148.
   see also land ordinances
home value, 40
hot tubs, 68
humid climates, 194
hybrid systems, 247, 262, 282, 291
hydrocaps, 216
hydroelectricity, 10, 21, 267, 268–269.
   see also microhydro systems
hydrofluorocarbons, 128–129

I
Icynene, 50–51
incentives: for energy conservation, 39–40,
   72; for heat pumps, 129; loans, 128,
   199; for solar electric systems, 23, 24,
   198–200, 219, 237; for solar hot air
   space heating systems, 141; for solar
   hot water systems, 82–83, 106, 147; for
   wind power, 263
induced draft water heaters, 77
installation: of microhydro systems, 275,
   290; of PV modules/arrays, 135–139;
   of solar electric systems, 221, 224,
   237–238; of solar hot air space heating

systems, 135–139; of solar hot water
   systems, 104–105; of wind power, 250,
   258, 260, 261–262; of wood stoves,
   158–159, 160
insulation: of ceilings/roofs, 53–57,
   121; of cooling ducts, 192; in different climates, 54–55; effect of moisture
   on, 141–142; as energy conservation,
   38–39; hiring for, 50–51; and older
   homes, 42; for passive cooling, 185; and
   passive solar retrofits, 117; recommendations for, 46–47, 49–50; in sunspaces,
   121; for walls, 50–53; of water heating
   systems, 65; in window shades, 184
Internet, 224, 225
inverters: buying, 229–232; and microhydro, 271; and solar electricity,
   203–204, 214, 218, 219; and wind
   turbines, 245, 261
invester-owned utilities (IOUs), 199

J
Jack Rabbit, 284–285

K
Keegan, Pat, 42

L
land ordinances, 83–84, 148, 240, 241,
   292
lead acid batteries. *see* batteries
lead sulfate, 217–218
leaks, plugging, 38, 39, 45, 46, 48–49, 185,
   192

LED lights, 67–68

lifestyle changes, 31, 33, 90, 176, 177–178

lighting: energy conversion of lightbulbs, 16; energy efficient, 32, 33–36, 66–68; and insulation, 55; and leaks by light-switches, 45, 49; and passive cooling, 176, 177; skylights, 59, 123–124; wattage requirements for, 222

loans, 128, 199

local economies, 29

M

magnetic declination, 113–114

maintenance: of batteries, 214–218, 234, 236; of masonry heaters, 171, 172; of microhydro systems, 283; of PV modules/arrays, 215; reduction of, 40; of sealed batteries, 236; of solar hot air systems, 140–141; and tankless water heaters, 80; of thermosiphon systems, 96; of wind turbines, 260–261; of wood stoves, 159, 160–161

marine deep-cycle batteries, 234

masonry heaters, 166–172, 171, 172

mass walls, 118–119, 121

mastic, 192

mechanical energy, 18–19

meters, 204–208, 210

microhydro systems (*see also* hydroelectricity): assessing energy needs of, 273–275; assessing potential of site, 275–280; and batteries, 281, 286, 287–288; choosing a turbine/generator, 283–285; choosing components for, 283–289; cost of, 269, 292; described, 267–268, 269; designing, 280–289; installation, 275, 290; location for, 281; maintenance of, 283; permits for, 289–290, 292; pros and cons of, 290–292; two types of, 270–271

microinverters, 203–204

microwaves, 70–71, 176

mildew, 54

mineral deposits, 147

mold, 54

monocrystalline solar cells, 227

N

natural conditioning, 174

natural gas, 10, 11, 12

neighborhood associations, 83–84, 148. *see also* land ordinances

net energy yield, 13

net metering, 207–208

new homes: and energy efficiency, 32–33, 42; and energy independence, 28; insulation of, 49–50; and passive cooling, 174; and passive solar, 110–111, 119; and sizing solar electricity systems, 221

The New Solar Electric Home (Davidson), 224

Newhall, Nancy, 73

noise, 231, 238, 254

nonrenewable energy, 10, 11, 13, 14–15, 16, 27, 242–243

North American Board of Certified Energy Practitioners (NABCEP), 237

nuclear energy, 10

O

off-grid systems: and batteries for, 234; and buying an inverter, 229–230; maintenance of solar, 215–218; and microhydro, 274; set up of solar, 210–214; sizing, 221; and wind power, 246, 262

oil, 10, 11, 242–243

oil shale, 13

ovens, 177

overhangs, 116, 117, 182, 183

ozone destruction, 129

P

paint/painting, 186

passive cooling: and airconditioners, 192–193; experience of, 173–174; and fans, 189–191, 193–194; how it works, 174–175; and insulation, 185; list of steps for, 195–196; and reducing external heat, 176, 179–186; and reducing internal heat sources, 175–176, 177–178; and ventilation, 188–189

passive solar: author's use of, 5, 109; cookers, 122; deciding if it's for you, 112–114, 124, 125–126; described, 110–112; economics of, 109, 125–126; hiring for, 124–125; sunspaces, 119–124; thermal storage walls, 118–119; and windows, 115–118

peak oil, 11

pellet stoves, 162, 164–166

permanent magnet alternator, 245, 284

permits: for microhydro systems, 289–290, 292; for solar electricity, 219, 237; for solar hot water space heating systems, 148; and sunspaces, 124

pets, 178

phantom loads, 70–72

Photovoltaics: Design and Installation (Weiss), 224

pipeline, 288

plexiglass, 58–59

pollution: from coal, 12; from ground-source heat pumps, 128–129; indoors, 161; from wood furnaces, 163; and wood stoves, 152, 156, 159, 161

polycrystalline solar cells, 227–228

power centers, 212–214

power lines, 289

power strips, 70, 71

power vented water heaters, 77

powerhouses, 286–287

progressive tube water heaters, 88–89

propylene glycol. *see* antifreeze

pump system water heaters, 96–100

PV laminate, 228, 229

PV modules/arrays: and battery size, 235; buying, 225–226; combined with wind power, 247; compared to microhydro, 291; in grid-connected systems with battery backup, 210; installing ground-mounted, 136; installing hot air, 135–139; installing roof-mounted, 136–137; and inverters, 230; maintenance of, 215; pictures of, 4, 23; placement for solar electric systems, 202; to run hot water system pump, 96, 145; types of, 226–229; for water heaters, 84, 91–94, 103–104

R

radiant barriers, 186, 187

reading lamps, 67

Reagan, Ronald, 82

Real Goods, 225

rebates, 199, 237

recycling, 29

renewable energy: benefits of, 25–26; disadvantages of, 26–27; early history of, 9–10; forms of, 14, 21, 22; future of, 18; ideas for, 28–29; lifestyle while using, 31, 33; myths about, 31; stockpiling of, 26–27

retrofits (*see also* passive solar): and appliances, 61–62, 68–72; and energy audits, 42–47; as energy conservation, 38–39; grants for, 1–2; hiring for, 47–48; insulation, 49–57; masonry heaters, 169, 170, 171; sealing cracks, 48–49; and showers, 65–66; and water heaters, 64–65; and windows, 57–61, 115, 116

rock storage bins, 133

roofs: installing hot air solar collectors, 136–137; insulation of, 56–57; and new solar collectors, 228–229; and passive cooling, 186; and solar hot water heaters, 90; and tree shade, 181

S

safety: and batteries, 215, 234–235; and heat pumps, 128; and solar electricity, 218; and solar hot air systems, 133; and water heaters, 77; and wood furnaces,

163, 164; and wood stoves, 152, 153, 157, 158, 160, 162

sand beds, 146

sealed batteries, 235–237

selective surfaces, 92

shade, 180–185, 192

showers/showerheads, 33–34, 65–66, 178

sizing: air conditioners, 192, 193; batteries, 235; masonry heaters, 170; solar electric systems, 220–223; solar hot water systems, 101–104; solar hot water systems for space heating, 147; wind turbines, 262; wood stoves, 157–158

skylights, 59, 123–124

sleepers, 56

smoke sticks, 45

solar cells, 23, 200–202, 226–229

solar collectors. *see* PV modules/arrays

solar electric systems (*see also* PV modules/arrays): benefits of, 238; buying batteries for, 232–235; buying inverters for, 229–232; buying PV modules for, 225–226; combined with wind power, 247; cost of, 32, 33; deciding which system is for you, 218–220; economics of, 197–200, 220, 238; finding a contractor, 237–238; further resources on, 224–225; grid-connected PV systems, 200–208; grid-connected systems with battery backup, 208–210; incentives for, 23, 24, 198–200, 219, 237; installation, 221, 224, 237–238; maintenance of, 214–218; off-grid systems, 210–214; power centers, 212–214; shopping for,

223–225; sizing, 220–223; supplying power to grid, 204, 205–208; and tree shade, 181; types of PV module for, 226–229

solar energy (*see also* passive solar): author's early passion for, 3–4; cost compared to conventional power, 22–24; financial incentives for, 23, 24; in food, 9–10; gained from windows, 60; varieties of, 22

solar hot air space heating systems: described, 131–132; economics of, 140–141; effectiveness of, 139–140; how they work, 132–133; incentives for, 141; installing, 135–139; shopping for, 141–142; types of, 133–135

solar hot water systems: barriers to, 83–84; batch heaters, 85–88, 89–90; choosing a system, 100–101, 107; described, 81; economics of, 105–107; hiring an installer, 104–105; history of, 81–84; incentives for, 82–83, 106, 147; installing your own, 105; lifestyle changes and, 90; progressive tube heaters, 88–89; PV modules for, 84, 91–94, 103–104; sizing, 101–104; for space heating, 13, 81, 143–148; for swimming pools, 102–103; types of, 84–85, 94–100

solar screens, 184–185

solar shades, 185

Solatron Technologies, 225

space heating systems (*see also* solar electric systems; solar hot air space heating systems): fireplaces, 149–150; furnaces and boilers, 62, 64; heat pumps, 62–64,

126–129; ideas for, 12–13; masonry heaters, 166–172; and passive solar, 110; pellet stoves, 164–166; from solar hot water, 13, 81, 143–148; wood furnaces, 162–164; wood stoves, 150–162

stack effect, 188

stand-alone system, 210–214

standby loss, 78

stereos, 178

storage tanks, 146

stoves, 177

sulfur, 12

sun, 21–22

sun-drenching, 116, 125

sunspaces, 119–124

surge power rating, 230–231

swimming pools, 102–103

T

tags, green, 264

tankless water heaters, 78–80, 81

tar sands, 13

tax credits, 106, 219, 263

televisions, 178

Tesla, Nikola, 242

thermoshutters, 117

thermosiphon systems, 94–96

tile stoves, 169. *see also* masonry heaters

tilt-up towers, 259–260

trackers, 23, 202

transpired air collectors, 134, 139, 141–142

trees, shade, 180–181, 182, 186

Trombe wall, 118–119

tubular skylights, 123–124
240-volt electricity, 232

U
UniSolar, 228–229
utility companies: and backfed electricity on grid, 204, 205–208; and cost of connection to grid, 198; and energy audits, 47; and energy conservation, 38–39; and green power, 263–264; and rebates, 237; and retrofitting, 48; and setting up for solar electricity, 219

V
vapor barriers, 54–55
vegetable growing, 122–123
ventilation, 188–189
voltage, 20

W
walls, 50–53, 118–119, 121
washing machines, 68, 177
waste heat, 175
water conservation, 33–34, 65–66, 101
water heating systems (*see also* solar hot water systems): as aid to space heating, 13; combined conventional/solar, 81, 86–89; cost of, 75; extending life of water heaters, 78; and heat pumps, 63, 128; how conventional water heaters work, 75–78; making water heaters efficient, 64–65; and passive cooling, 176, 177; tankless, 78–80; wood furnaces, 162, 164

Water Miser caps, 216
water-source heat pumps, 127
watts, 20–21
weatherizing homes, 46–47. *see also* insulation
Whisper Aire fan, 190–191
wind power: assessing wind resource, 247–249; buying a tower, 255–259, 259–260; in cities and towns, 240–241; combined with solar, 247; compared to microhydro, 291; cost of, 262–263; deciding turbine location, 249, 261; finding a supplier/installer, 260; history of, 241–243; how turbines work, 243, 245; incentives for, 263; installation, 250, 258, 260, 261–262; maintenance, 260–261; noise from, 254; picture of turbine tower, 28; selecting turbine and tower, 249–255; three options of, 246–247; types of turbine, 244–245
windmills, 241, 242
windows: and passive cooling, 185–186; and passive solar retrofits, 115–118; shades for, 183–185; in sunspaces, 122; upgrading, 57–61
wood heat, 149–150, 162–164
wood stoves: installing, 158–159, 160; masonry heaters, 166–172; pros and cons of, 158–162; shopping for, 154–157; sizing, 157–158; types of, 64, 150–153

Z
zoning ordinances, 83–84, 148. *see also* land ordinances

About the Author

Dan Chiras is an author, lecturer, consultant, and educator. He has spent more than 30 years of his life studying renewable energy and energy efficiency and other aspects of sustainable design and has been applying what he has learned in these areas to his homes and businesses.

Dan is founder and director of The Evergreen Institute's Center for Renewable Energy and Green Building, located in east-central Missouri (evergreeninstitute. org), where he teaches classes on home energy efficiency, solar electricity, wind energy, electric wiring, the National Electric Code, passive solar heating and cooling, green building, natural building, and natural plasters.

Dan has been a visiting professor at Colorado College for more than a decade. Here he teaches courses on renewable energy, sustainable development, green building, and ecological design.

Dan has published nearly 300 articles on environmental issues, sustainable development, green building, and renewable energy in a variety of magazines, journals, newspapers, and encyclopedias. His articles routinely appear in *Mother Earth News, Home Power, Solar Today,* and *Natural Home.*

Dan has published numerous books — 29 as of December 2010 — including *Power from the Wind; Power from the Sun; Wind Power Basics; Solar Electricity Basics; Green Home Improvement; The Solar House: Passive Solar Heating and Cooling; The Natural House; The New Ecological Home; Green Transportation Basics; The Natural Plaster; Superbia! 31 Ways to Create Sustainable Neighborhoods;* and

EcoKids: Raising Children Who Care for the Earth. In 2006, he published his first novel, *Here Stands Marshall.*

Dan consults on residential passive solar heating and cooling design and green building throughout the United States, Canada, and Central America through his company, Sustainable Systems Design, Inc. Dan is well on his way to converting The Evergreen Institute's educational center to a net zero-energy facility.

If you have enjoyed *The Homeowner's Guide to Renewable Energy*,
you might also enjoy other

BOOKS TO BUILD A NEW SOCIETY

Our books provide positive solutions for people who want to
make a difference. We specialize in:

Sustainable Living • Green Building • Peak Oil • Renewable Energy
Environment & Economy • Natural Building & Appropriate Technology
Progressive Leadership • Resistance and Community
Educational & Parenting Resources

New Society Publishers

ENVIRONMENTAL BENEFITS STATEMENT

New Society Publishers has chosen to produce this book on recycled paper made with
100% post consumer waste, processed chlorine free, and old growth free.

For every 5,000 books printed, New Society saves the following resources:[1]

45	Trees
4,087	Pounds of Solid Waste
4,496	Gallons of Water
5,865	Kilowatt Hours of Electricity
7,429	Pounds of Greenhouse Gases
32	Pounds of HAPs, VOCs, and AOX Combined
11	Cubic Yards of Landfill Space

---

[1]Environmental benefits are calculated based on research done by the Environmental Defense Fund and
other members of the Paper Task Force who study the environmental impacts of the paper industry.

For a full list of NSP's titles, please call 1-800-567-6772 *or check out our website* at:

www.newsociety.com

NEW SOCIETY PUBLISHERS
Deep Green for over 30 years